Remington Autoloading & Pump-Action Rifles

A history of the centerfire Models 760, 740, 742, 7400 & 7600

Eugene Myszkowski

Excalibur Publications

Dedication

To Jan, for putting up with my latest quest. She took my bringing home an engraved F grade with gold inlaid game scenes in stride. One could not ask for a better partner to share the journey.

Other books by Eugene Myszkowski,
also published by Excalibur Publications:

The Remington-Lee Rifle
The Winchester-Lee Rifle

Remington Arms Company Inc. has given permission to use its tradename and trademarks, as well as illustrations from its catalogs and brochures.

 **Excalibur Publications**

PO Box 35369, Tucson, AZ 85740-5369
voice: (520) 575-9057
fax: (520) 575-9068
email: excalibureditor@earthlink.net

ISBN # 1-880677-20-2
First printing — May, 2002

Contents

On the cover: First production rifles presented to Remington President C.K. Davis. Both are F grade engraved with gold inlaid game scenes. The upper rifle is a Model 760, serial number 1001; the lower is a Model 740, serial number 1001.

Chapter 1
The Beginning

Sometimes the most important person in bringing a new product successfully to market is not the inventor, the production engineer or the salesman but the "boss." In the case of Remington Arms Company's autoloader Model 740 and pump action Model 760, company President Charles Krum Davis kept the focus on the goal of a successful gun design. It was apparent when Davis took over in 1933 that the Model 8 autoloader and the Model 14 pump action rifles were dated and expensive to manufacture. The rifles needed, at a minimum, cosmetic updating. The Model 141 pump action rifle appeared in 1935 and the Model 81 autoloader rifle in 1936 as stopgap measures.

Under his leadership, the entire line of Remington firearms was upgraded and later replaced with new designs, many following the "family of guns" concept.

Crawford C. Loomis and his design team was authorized to begin work on the replacement for the Model 141 pump action rifle—the Model 49 pump action rifle —on January 20, 1936. The first major change was to substitute a box magazine for the tube magazine on April 10, 1938. The Remington Model 49 was originally designed to handle high power cartridges in the .375 H&H Magnum class. A Model 49 was completed in .30-06 and demonstrated on October 5, 1938; however, the receiver depth, length and width were criticized by the design review committee as being too large. A discussion followed concerning the feasibility of producing two receivers: one for magnums, and the other for cartridges in the .30-06 class. The result was that the Model 49 was dropped and work was begun in early 1939 on a new model pump action rifle for the .30-06 cartridge. This new

C.K. Davis, President of Remington Arms Company Inc., 1933 – 1954

4

model was given the engineering studies designation, Model 760, on June 13, 1940.

In 1939 George E. Pinckney, Manager of the Product Sales Division, took a trip through the Midwest and along the Pacific coast. His subsequent report indicated a large consumer demand for a high power slide action rifle capable of handling cartridges in the .30-06 class.

A progress report on the Model 760 was given July 25, 1941 by R. A. A. Hentschel of Remington's Research Division and a sample rifle was demonstrated for the Sales Department on September 19, 1941. In addition, a progress report on the autoloader Model 740 was given almost a month later. However, all work on both rifles was put on hold after the Japanese bombed Pearl Harbor on December 7, 1941.

The Remington factory already was producing the M1903 Springfield at the request of the British Purchasing Commission. In June 1940, shortly after the British experienced huge losses at Dunkirk, the British Purchasing Commission asked Remington to send a representative to Washington, D.C. to discuss rifle production. Donald F. Carpenter, then Remington's Director of Manufacturing, met with high-ranking members of the British Purchasing Commission and US civilian and military officials. It was decided that the government would ship Remington the stored Rock Island Arsenal equipment for manufacturing the M1903 Springfield and the first rifles

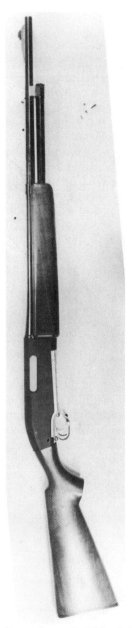

Experimental Model 740, caliber .30-06, circa 1944. The aluminum alloy receiver is large enough to accommodate the .300 H&H Magnum cartridge. *Photo: Courtesy of the Remington Arms Co.*

under the British contract were produced just prior to the Pearl Harbor attack on December 7, 1941.

After Pearl Harbor, the US took over the British contract and doubled the production schedule to 2,000 rifles per day. In order to increase production to accommodate that schedule, it was necessary to simplify the M1903 rifle. L. Ray Crittendon of Du Pont's Engineering Department was assigned the job of simplification and within a few months had a model of a revised rifle. This model, incorporating many stampings which replaced machined parts, became the M1903A3 Springfield. Some design work at Remington resumed in late 1943 on both Models 740 and 760 and three action sizes were developed for the pump action rifle: Model 760 for large cartridges in the .300 H&H Magnum and .30-06 class; Model 761 for medium cartridges in the .35 Remington class; and Model 762 for small cartridges in the .22 Hornet and .30 Carbine class. A caliber committee met in 1944 and proposed the following new cartridges for introduction in 1948 in all new Remington rifles; however, no action was taken:

- .224 Remington (to replace .219 Zipper and .220 Swift)
- .258 Remington (to replace .257 Roberts)
- .276 Remington (a new cartridge, based on the .300 Savage case)
- .280 Remington (to replace 7mm Mauser)
- .284 Remington Magnum (to replace .270 Winchester, based on .300 H&H Magnum case)
- .308 Remington Magnum (to replace .300 H&H Magnum)
- .350 Remington (to replace .35 Remington)
- .359 Remington Magnum (to replace .375 H&H Magnum)

On April 20, 1944 a sample Model 760 in caliber .30-06 was fired 3,000 times to test its reliability, but further work to ease its operation was needed. A formal proposal for pump action Models 760 and 761 development, testing and preparation for manufacture was approved May 5, 1944. The companion autoloader Models 740 and 741 proposal was requested July 12, 1944.

On August 23, 1944 Davis reported to the Board of Directors on the future direction of Remington. He discussed new designs, such as the Model 740 and Model 760, and noted that all rifles used sound engineering principles and the latest post-war production methods. A study of design and production procedures was conducted by Remington staff, assisted by the Engineering Department of the Du Pont Corporation and in consultation with members of the General Motors Corporation. It was Crittendon, of Du Pont's Engineering Department, who had incorporated modern production methods in his design of the M1903A3 Springfield.

There were two gun design teams working by 1947: the Remington Technical Division in Ilion, NY and the Du Pont Engineering Department in Wilmington, Delaware. The Remington Technical Division was working on its original designs: the autoloader Models

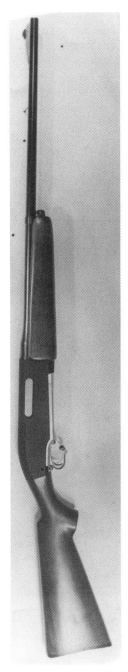

Experimental Model 760, caliber .30-06, circa 1944. *Photo: Courtesy of the Remington Arms Co.*

740 and 741 and the pump action Models 760 and 761. The small frame Model 762 had been dropped sometime previously.

Crittendon of Du Pont had begun work on May 20, 1947 on entirely new autoloader and pump action rifle designs. They were given the engineering designations Model 742 and Model 762 on October 9, 1947. [Note: The use of the same model numbers as previous rifles caused this author considerable confusion, as the various designs were discussed in committee.]

These new designs were based on the size and appearance of the 28 and .410 gauge Model 851, later called the Model 11-48, shotgun receiver. The Model 851/11-48 shotgun, which had a recoiling barrel, was familiar to Crittendon as he, William Gail Jr., Ellis Hailston and C. R. Johnson were the original design team members. The rifle receiver's major changes were a stationary barrel, which locks into the receiver by means of a barrel extension and a multiple lug bolt. It was estimated that 35 to 40 of the 80 parts found in the Model 851/11-48 shotgun would, with minimal revision, be common to a rifle.

The bolt locking mechanism was the subject of numerous tests during 1947. A three row multiple locking lug bolt was discussed on October 1 and tested for strength on October 30 and 31. It successfully withstood 10 proof loads. The proof load powder charges were then increased over 20 percent and the bolt still held.

The two design teams, at Ilion and Wilmington, inevitably

7

had a turf war, and President Davis had to resolve the conflict. In a meeting with the principals from both design teams on November 11, he emphasized that his responsibility was looking out for the stockholders' interests, which meant producing a gun of the best design, regardless of who designed it. The Remington Technical Division was eventually given responsibility for the gun design with the Du Pont Engineering Department providing facilities and staff. Thus Crittendon, Gail Jr. and other Du Pont engineers became part of the gun design team.

By 1949 the autoloader Model 742 and pump action Model 762 designs were selected and endurance trials were well under way. The bolt was changed to a four row multiple locking lug design in 1950. By mid-1950 formal trial and testing was concentrated on the pump action rifle to prepare it for production. Field trials of the Model 760, as it was now called, were successful and only the magazine presented continuing problems. The autoloader, now renamed the Model 740, still had not passed extended function and endurance trials.

Chapter 2

Remington Model 760

The engineering designation Model 762 for the new pump action rifle was formally changed on May 22, 1950 to the Remington Model 760. One of the reasons for changing the model number was to eliminate any confusion by the public with the 7.62mm Russian cartridge. Later it was decided to use the *Gamemaster* trade name to tap into the advertising value, goodwill and reputation of the Remington Model 141 pump action rifle.

Four pilot Model 760s in caliber .30-06 were sent on August 10 to various Remington District managers for demonstrations, evaluation and field trials. The overall consensus was favorable, but the fore-end, a modified Model 141 style, was criticized for being too thin, fragile and having sharp edges. Extraction problems, hard unlocking, a flimsy ejector port cover and the need to push up on the magazine in order to release the catch also were criticisms.

Six more pilot Model 760s were subjected to expanded endurance and function trials during the late summer and fall of 1950, which uncovered problems with extractors breaking during proof firing, hard unlocking and magazine feed jams. All these difficulties were eventually resolved.

Remington, in a memo January 18, 1951, proposed an announcement date of July 1. The company wanted to have

The initial 1952 announcement flier for the Model 760, front and back.
Courtesy of Jack Heath.

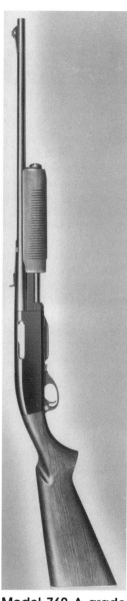

Model 760 A grade rifle, with low comb stock. *Courtesy of Remington Arms Company.*

Preproduction Model 760 ADL/BDL grade rifle with figured wood that is usually found on higher grade rifles. *Courtesy of Jack Heath.*

Model 760 A grade rifle with higher comb "all purpose" stock introduced in 1958. The base mounted step and windage adjustable rear sight was introduced in 1960. *Courtesy of Remington Arms Company.*

1,800 rifles in the warehouse ready for sale and 2,100 more in production. This schedule quickly slipped six months to January 3, 1952, when the Model 760 was formally introduced to the public in the 1952 Remington consumer catalog. Dealers received the new pricing schedule on February 6, 1952.

An artillery style bolt with multiple locking lugs and an encased bolt head that completely supported and enclosed the cartridge case, (similar to the bolt action Models 721 and 722), twin action bars and a detachable box magazine were all showcased in the initial catalog description. The new rifle was offered in .30-06, .300 Savage and .35 Remington, and in grades A, B, D and F.

The sporting press received trial rifles in late 1951, and writers like Warren Page of *Field*

Model 760 pump action rifle hang tag circa 1958. *Author's collection.*

and Stream, Julian Hatcher of the National Rifle Association, Elmer Keith of the *American Rifleman,* Bert Popowski of the *Outdoorsman,* and Edson Hall of the *Outdoor Sportsman* were favorably impressed. Most of the gun writers commented on a need to have the receiver drilled and tapped for scope mounts and for a higher comb stock. Elmer Keith sent his rifle out on a successful elk hunt and suggested that the rifle also should be chambered for either the .333 O.K.H. or the .35 Whelen for truly big game.

In early May 1952, Remington experimented with a Remington Model 760 chambered for the .35 Whelen cartridge. On his own Elmer Keith also rechambered a caliber .35 Remington Model 760 to .35 Whelen. However, the caliber was not added to the line because of problems with maintaining headspace.

In 1953 Remington designer

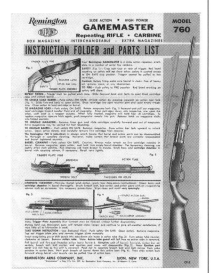

Early instruction folder for the Model 760 with detailed instructions for disassembly of the rifle. *Courtesy of Jack Heath.*

11

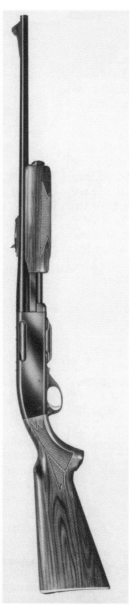

Model 760 ADL grade rifle with "all purpose" stock introduced in 1958. The base mounted step and windage adjustable rear sight was introduced in 1960. *Courtesy of Remington Arms Company.*

Model 760 Standard grade rifle, with new pressed checkering pattern, introduced in 1964. *Courtesy of Remington Arms Company.*

Model 760 Standard grade rifle with 1968 revisions to fore-end assembly and pressed skip-line checkering. *Courtesy of Remington Arms Company.*

and engineer Wayne Leek examined the feasibility of offering extra barrels for the Model 760 so that a shooter might have different calibers in the same rifle. The design of the barrel to action attachment allowed a relative easy interchange, but it was not practical as the screw thread arrangement of the multiple locking lugs made headspacing an intricate and delicate operation. The question of interchangeable barrels resurfaced in 1954 as a number of gunsmiths inquired about extra barrels.

It was raised again in January 1966 when Dan Cotterman in that month's issue of *Gun World* published a field test article entitled, "REMINGTON'S MODEL 760 - This is known as a butterfly gun; with changee-changee barrels for numerous calibers." Dan checked with Remington prior to

Initial 1960 announcement flyer for the Model 760 C carbine. *Courtesy of Roy Marcot.*

publishing the article and included their disclaimer, as well as detailed instructions on how to change the barrels. The article caused some consternation at Remington, whose policy was never to sell Model 760 barrels to the public since only Remington could install replacement barrels.

After some discussion, the left receiver panel stamping was stamped *Remington* over the serial number in the lower middle, while *Gamemaster* over Model 760 was stamped between the fire control pins. The two line barrel legend was stamped on the left side of the barrel above the fore-end:
REMINGTON ARMS CO. INC., ILION, N.Y MADE IN U. S. A. PATENT NO. 2,473,373 OTHERS PENDING

The caliber was stamped after the barrel legend.

The standard A grade's plain low comb stock had a lacquer finish, a 13-1/2 inch length of pull, a 1-3/4inch drop at the comb and a 2-3/4 inch drop at the heel. The fore-end was a "semi beavertail with an extension," having 28 vertical grooves on each side. An aluminum shotgun-style buttplate, with bright finish on the edges, was fitted. The ramp front sight had a white metal bead, while the step-adjustable buckhorn rear sight was dovetailed in the barrel. The receiver was not drilled and tapped for scope mounts or a receiver sight.

1952

The first Model 760s were offered in caliber .30-06 and were made available in February 1952. The first .300 Savage rifles were shipped in April 1952, and the first .35 Remington rifles were shipped in July of that year. Almost all 1952 production rifles were A grade with a low comb stock. Priced at $104.40, 1952 sales totaled 63,735 rifles (41,417 in caliber .30-06; 14,431 in .300 Savage; and 7,887 in .35 Remington).

The first production rifle, serial number 1001, was given to Remington Arms Company President C. K. Davis. It was an "F" grade with gold inlaid game scenes and is one of the rifles illustrated on the cover.

Work on an ADL "Deluxe" grade Model 760 was started in 1951, however, it wasn't until June 10, 1952 that the design was finalized. Rifle serial number 17418, caliber .30-06, was approved with a checkered high comb buttstock, which had a length of pull of 13-1/2 inches, a drop at the comb of 1-1/4 inches, and a drop at the heel of 2 inches. The fore-end checkering pattern is called by today's collectors "5 Diamond" (not a Remington designation). The action tube end cap was drilled and tapped for sling swivels and the receiver was drilled and tapped for scope mounts. A low comb option was made available. Later a BDL "Deluxe Special" grade with "selected" wood was approved for production. The ADL grade was announced on December 31, 1952, the BDL grade a short time later and deliveries for both models began in March 1953. The price was $119.95 for the ADL grade and $139.70 for the BDL grade.

Normally there were no markings on the rifle to distinguish between early ADL and BDL grades, however, a few BDL grade rifles have been observed with the grade engraved after the model number. The original shipping box end label had the grade noted and provides positive proof of an early BDL grade rifle. Many observed ADL grade rifles have nicely figured wood. Sometimes, due to a flaw, BDL grade wood was downgraded and installed on ADL grade rifles.

Instruction folder for Model 760 C carbine with disassembly instructions. The quoted 6-1/2 pound weight is incorrect. *Courtesy of Roy Marcot.*

TAKE DOWN PIN HOLES

SAFETY   MAGAZINE LATCH

OPERATION of CARBINE

Your Remington Model 760 in this Carbine model with its short barrel is ideal as a brush hunting or scabbard firearm. This five (5) shot GAMEMASTER Carbine weighs only 6½ lbs. and is calibered for big game, long range shooting. The four (4) shot, box magazine, as in the standard Model 760 rifle, may be easily removed for quicker reloading. Extra magazines are available from your dealer. The operation and sighting instructions are the same as outlined in the standard M/760 instruction folder.

TAKE DOWN of CARBINE

And necessary cleaning and care are also as directed in the standard Model 760 Instruction Folder — RD 5551. However, the take-down pin holes in the action tube of the Carbine model are located to the rear of the fore-end, when the action is closed. See illustration above. Note also in Carbine model, small bolt carrier spring to rear of bolt. See Sectional View below.

REPLACEMENT PARTS, if needed, can be identified from Sectional View.

REAR SIGHT ASSEMBLY
REAR SIGHT COLLAR
REAR SIGHT WINDAGE SCREW
REAR SIGHT EYEPIECE
REAR SIGHT LEAF
REAR SIGHT BASE
BREECH BOLT ASSEMBLY
BARREL ASSEMBLY
FORE END CAP SCREW
BOLT CARRIER
TAKE DOWN
ACTION TUBE FORE END
FORE END
FORE END
SPRING
PIN HOLES
ASSEMBLY
TUBE YOKE
FORE END CAP
FORE END PLUG
FORE END TUBE ASSEMBLY
M-760-C SLIDE ACTION CENTER FIRE CARBINE

NOTE: Parts not labeled — same as standard Model 760.

SECTIONAL VIEW of CARBINE

Model 760 Standard grade rifle with 1973's minor change in buttstock checkering pattern. The sliding ramp style rear sight was introduced in 1975. *Courtesy of Remington Arms Company.*

Model 760 C grade carbine introduced in 1960. The buckhorn rear sight is incorrect as production carbines were equipped with the new base-mounted step and windage adjustable rear sight. *Courtesy of Jack Heath.*

Model 760 prototype CDL grade carbine as illustrated in catalog but not put into production. The fore-end and sling swivels are from the ADL grade rifle. *Courtesy of Jack Heath.*

The factory offered to upgrade A grade rifles to ADL grade for $20.00 and also would drill and tap for scope mounts for $ 6.50 or fit sling swivels for $5.00. Any rifles returned to the factory for repairs and/or upgrades would be stamped **3**, a date code and the inspector character on the left side of the barrel ahead of the assembly date code and character.

A number of other changes were made during 1952 and 1953. The action tube cap originally had two coin slots — these were eliminated in a June 1952 revision. The fore-end was slightly redesigned with a taper, widening to the rear, to eliminate cracking in the rear portion.

Several changes were made to the action bar and action bar lock to prevent "premature opening," and an auxiliary inner hammer plunger spring was added to increase the force of the firing pin blow. The bolt carrier assembly also was revised.

1953

Caliber .270 Winchester was added to the line in the March 31, 1953 Remington catalog. In addition, ADL and BDL grades were added to the catalog in place of the B grade. Total 1953 sales for all calibers and grades were 81,428 rifles. Of these 37,553 were A grade caliber .30-06 rifles. 1953 ADL and BDL grade sales were 11,822 rifles in all calibers.

In mid-1953, around serial number 100,000, the radius of the top of the Model 760 receiver was increased. Consumer complaints about "rattle" of the ejection port cover and fore-end triggered research to eliminate the problem. The barrel assembly was changed from a full conical feed ramp to a feed ramp only in the lower quarter of the barrel for increased support of the cartridge case.

1954

Caliber .257 Roberts was added to the line on July 6, 1954. Total Model 760 sales during 1954 were 85,065. The ejection port cover was redesigned and the ejection port relief cut in the barrel extension revised to a fillet cut.

1955

Total sales of the Remington Model 760 dropped to 30,099 because of the introduction of the Remington Model 740 autoloader. The ejection port cover on the Model 760 was changed from steel to a black nylon resin called "Zytel." There was some discussion of introducing a new caliber — the .266 Remington in the Model 760, but no action taken.

1956

The two line barrel legend stamped on the left side of the barrel above the fore-end had the patent number changed on September 26, 1956 to:
REMINGTON ARMS CO. INC., ILION, N.Y MADE IN U. S. A. PATENT NO. 2,685,754 OTHERS PENDING

The following calibers were added to the line: .244 Remington on October 9, 1956; and .308 Winchester on December 27, 1956.

1957

The .222 Remington was added on June 15, 1957. The .257 Roberts was dropped and the catalog noted that sales are "subject to stock on hand."

16

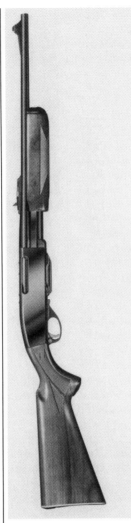

Model 760 CDL grade carbine as produced. *Courtesy of Remington Arms Company.*

Model 760 Standard grade carbine introduced 1964. *Courtesy of Remington Arms Company.*

Model 760 Standard grade carbine with pressed skip-line checkering pattern introduced in 1968. *Courtesy of Remington Arms Company.*

1958

The two line barrel legend stamped on the left side of the barrel above the fore-end was changed to reflect granting of additional patents: REMINGTON ARMS CO. INC., ILION, N.Y MADE IN U. S. A. PATENT No. 2,685,754 AND OTHERS

Caliber .280 Remington was added in the 1958 Remington catalog. The .300 Savage cartridge was dropped and the catalog noted that .257 Roberts and .300 Savage sales are "subject to stock on hand." The BDL grade also was dropped. The catalog continued to offer the high and low comb option; however, internal factory memorandums note that an "all purpose" stock, substantially similar to the high comb option, "will be fitted to the A and ADL grades." The "all purpose" stocks have a 13-1/2 inch length of pull, a 1-3/4 inch drop at the comb and a 2-1/4 inch drop at the heel. The 1958 and later production records separate the "all purpose" stocks from the high and low comb options.

1959

Calibers .222 Remington, .244 Remington and .35 Remington were dropped on October 29, 1959. A carbine model of the Model 760 was under development.

1960

The Remington Model 760 C carbine was introduced in the 1960 catalog with calibers .30-06 available in March, .270 Winchester in May, and the .280 Remington in July. The .308 Winchester was added July 15, 1960. 760 CARBINE was

Mid 1970's standup cardboard poster advertising the Models 742 and 760 rifles. The painting is by Robert Kuhn. *Author's collection.*

stamped in large letters after the caliber designation on the barrel. The carbine fore-end and action tube differed from the rifle model in that the fore-end covers the action tube when fully retracted. This required that the disassembly holes in the action tube be moved from the front to the rear. An anti-rattle neoprene "O" ring was fitted in the fore-end of the carbine. This was so successful that it was quickly added to the rifles. The carbine barrel is 18-1/2 inches long and the overall length is 38-1/2 inches. The 1960 catalog and brochures gave the carbine weight as 6-1/2 pounds, this was increased to 7 pounds in the 1961 catalog and to 7-1/4 pounds in the 1962 catalog. All Model 760 carbines examined by the author have weighed around 7-1/4 pounds.

The carbine illustrated in the Remington catalog had the older style buckhorn rear sight dovetailed in the barrel; however, the catalog description

BDL grade Model 760 introduced in 1966 with a stepped receiver and pressed basketweave checkering on the Monte Carlo buttstock and flat fore-end. *Courtesy of Remington Arms Company.*

was for the new rear sight.

The catalog also noted that .222 Remington, .244 Remington, .257 Roberts and the .300 Savage chamberings are "subject to stock on hand."

A new rear sight for rifles and carbines was introduced in 1960. It was step adjustable for elevation and adjustable for windage by a screw. The rear sight was attached to the barrel by screws and set on a base to be compatible with the sight line of the high comb stock. The ramp front sight was changed to a flat faced gold bead.

1961

The Model 760 CDL grade carbine was announced on January 1, 1961 with deliveries beginning February of that year. The buttstock, with rear sling swivel, was identical to the Model 760 ADL rifle while the receiver and 18-1/2 inch barrel were identical to the Model 760 C. The fore-end, a shorter version of the Model 760 ADL in shape and checkering, used the same action tube as the Model 760 C. The upper sling swivel was a .655 inch Judd barrel band mounted 5 inches from the muzzle.

The Model 760 C carbine catalog illustration remained the incorrect version.

1962

The catalog illustration for the Model 760 CDL grade carbine had the same fore-end, action tube, action tube cap with upper sling swivel and rear sight as the ADL grade rifle. [Note: Thus far, a Model 760 CDL carbine, as illustrated in the catalog, has not been observed.]

The 1962 catalog description

19

noted that Model 760s were consistent winners in International Running Deer Matches. Participants used the Model 760 in .222 Remington as the fast action of the rifle and the low recoil of the cartridge allowed rapid repeat shots. This claim continued through Remington's 1966 catalog.

1963

No changes were noted.

The Model 760 CDL illustration remained the "incorrect" version.

1964

A single "customized" Standard grade rifle and carbine with the new "pressed checkering" replaced A, ADL, C and CDL grades. The Standard grade stock specifications were length of pull 13-1/2 inches, drop at comb 1-3/4 inches, and drop at heel 2-1/4 inches.

Caliber .35 Remington was added to the carbine line in January. Caliber .223 Remington (5.56 mm) was added to the rifle line in May 1964.

1965

No changes were noted.

1966

The Model 760 BDL Deluxe was offered in calibers .30-06, .270 Winchester and .308 Winchester. The receiver had a distinct step up at the rear and the Monte Carlo buttstock, with a choice of right or left hand cheek piece, as well as the flat fore-end, had a basket weave checkering pattern. The left hand models had the safety reversed; however, ejection is still to the right side. The BDL grade stocks had a 13-5/16 inch length of pull, drop at comb was

1966 150ᵗʰ Anniversary standup cardboard poster advertising the Standard grade Model 760. The painting, Tolling up Antelope, is by Robert Kuhn. *Author's collection.*

1-5/8 inch and drop at heel was 2 -1/2 inches. The action tube followed the pattern of the carbine and did not extend past the fore-end. The takedown holes were in the rear of the tube.

All models now had a DuPont developed RK-W gloss wood finish.

The Peters Sporting ammunition catalog listed that the caliber .35 Remington was available in the Model 760 BDL grade; however, none were made.

The Model 760 carbine line dropped the .270 Winchester and the .280 Remington calibers.

A 150th Anniversary Model, in caliber .30-06, was offered. See the chapter on Commemorative Rifles for more details on this model.

A special run of 200 Model 760 rifles, in caliber .222 Remington, was produced for competition in the International Running Deer Matches. The author has not observed any of these Model 760s. It is likely that Remington supplied the Standard grade rifle and any changes in sights or stocks were made by the participants in these matches.

1967

The barrel legend stamped on the left side of the barrel above the fore-end was changed to three lines to include additional patents:
REMINGTON ARMS CO. INC., ILION, N.Y MADE IN U.S.A. PATENT NO. 2,685,754 - 2,585,195 2,474,313 – 2,675,638 – 2,473,313 AND OTHERS

As of July 5, 1967 the Standard grade rifle was changed to the carbine design action tube and fore-end.

Existing action tubes and fore-ends continued to be used until supplies were exhausted.

1968

The Standard grade rifle and carbine had a restyled buttstock and fore-end with pressed skip-line checkering. The aluminum buttplate was changed to black plastic and it, as well as the black fore-end tip and grip cap, now had white line spacers. The stocks had a 13-5/16 inch length of pull, 1-5/8 inch drop at the comb and 2-1/8 inch drop at the heel.

Caliber .243 Winchester was added to the rifle line while the .280 Remington was dropped from rifle line. The .35 Remington caliber was dropped from both the rifle and carbine lines.

The Model 760 serial number sequence was changed on November 26, 1968 as a result of the 1968 Gun Control Act requiring that no two guns from the same manufacturer have the same serial number. The initial Model 760 serial number sequence began at 1001 and ended at 549773. The new serial number sequence, now shared with the Model 742 autoloader rifle, began at A6900000.

1969

Caliber 6mm Remington was added to the rifle line and caliber .223 Remington was dropped from the catalog listing. The caliber .223 Remington magazine had feeding problems and sales did not justify the cost of redesigning the magazine. Vibra-honing for a mirror smooth interior finish was advertised.

The FBI purchased 850

Model 760 carbines in .308 Winchester.

The BDL grade was changed to the "Custom Deluxe".

1970

No changes were noted.

1971

No changes were noted.

Two Model 760s in caliber .25-06 Remington were tested, but the caliber was not added to the line

1972

No changes were noted

1973

A minor change was made in the Standard grade checkering pattern on the Model 760.

1974

The barrel legend stamped on the left side of the barrel above the fore-end was changed back to two lines:
REMINGTON ARMS CO. INC., ILION, N.Y

MADE IN U.S.A. PAT. NUMBER - 2,473,373

1975

The rear sight on the Model 760 was changed to the new sliding ramp style.

1976

A Bicentennial Model 760, in caliber .30-06, was offered in a special brochure. (See the chapter on Commemorative Rifles for more details.)

1977

No changes were noted

1978

No changes were noted.

1979

Caliber .35 Remington was reintroduced on a limited production basis for the Model 760, and later advertised with its own flyer.

1980

The Remington Model 760 was discontinued December 31,

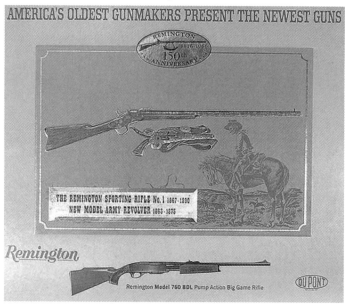

1966 150th Anniversary standup cardboard poster advertising the BDL grade Model 760. The painting of the Sporting Rifle #1 and New Model Army Revolver is by James K. Laing. *Author's collection.*

1980 with the introduction of the Model Six and Model 7600. However, sales continued well into 1981 to clean out warehouse stock.

The total production of the Model 760, including rifles, carbines, commemoratives and 82 engraved arms, was 1,034,438 firearms. A percentage of sales by caliber included:

- .30-06 - 63%
- .270 Win. – 16%,
- .308 Win. – 8%
- .300 Sav. – 4%
- .35 Rem. – 3%
- .243 Win. – 3%
- 6mm Rem. – 1%
- .280 Rem. - less than 1%
- .257 Roberts - less than 1%
- .244 Rem. - less than 1%
- .222 Rem. - less than 1%
- .223 Rem. - less than 1%

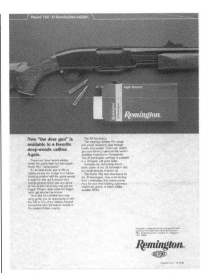

1979 announcement flyer reintroducing the caliber.35 Remington Model 760. *Author's collection.*

23

Table 1
Rifles Produced by Caliber and Grade

Caliber	Grade	Years Made	Order Numbers	Production	Total by Caliber
.30-06	A-Lo	1952-57		147,437	
	ADL*	1952-57		26,392	
	BDL*	1952-57		429	
	A	1958-63	5888	40,085	
	ADL	1958-63	5890	10,826	
	Standard	1964-67	9864	40,374	
	150th Ann.	1966		4,610	
	BDL**	1966-80	9692/9694	70,694	
	Standard	1968-80	9730	259,915	
	Bicentennial	1976	7614	3,804	
					604,566
.270 Win.	A-Lo	1953-57		43,635	
	ADL*	1953-57		12,906	
	BDL*	1953-57		142	
	A	1958-63	5870	7,714	
	ADL	1958-63	5874	2,186	
	Standard	1964-67	9680	9,355	
	BDL**	1966-80	9708/9712	22,431	
	Standard	1968-80	9678	61,030	
					159,399
.308 Win.	A-Lo	1957		2,964	
	ADL*	1957		945	
	BDL*	1957		11	
	A	1958-63	5896	5,755	
	ADL	1958-63	5900***	1,514	
	Standard	1964-67	9686	7,794	
	BDL**	1966-80	9706/9710	11,869	
	Standard	1968-80	9732	38,409	
					69,261
.300 Sav.	A-Lo	1952-57		37,797	
	ADL*	1953-57		3,909	
	BDL*	1953-57		38	
	A	1958		7	
					41,751

.35 Rem.	A-Lo	1952-57		19,818	
	ADL*	1953-57		1,922	
	BDL*	1953-57		27	
	A	1958-63	5910	782	
	ADL	1958-63	5912	120	
	Standard	1964-67	9688	1,200	
	Standard	1979-80	9733	11,932	
					35,801
.243 Win.	Standard	1968-80	9724	31,325	
					31,325
6mm Rem.	Standard	1969-80	5900***	10,519	
					10,519
.257 Rem.	A-Lo	1954-57		4,744	
	ADL*	1954-57		2,025	
	BDL*	1954-57		24	
	A	1958		4	
					6,797
.280 Rem	A	1958-63	5882	1,599	
	ADL	1958-63	5884	699	
	Standard	1964-67	9682	804	
					3,102
.244 Rem.	A-Lo	1956-59		2,101	
	ADL*	1956-59		970	
	BDL*	1956-59		9	
					3,080
.222 Rem.	A-Lo	1957		2,033	
	ADL*	1957		947	
	BDL*	1957		21	
	A	1958-66		304	
	ADL	1958		1	
					3,306
.223 Rem.	Standard	1964-69	9790	2,723	
					2,723
	D and F	1952-80		82	
					82
Total rifle production					971,712

Notes: * Includes both low and high comb options.
              ** Includes right and left handed options.
              *** Remington used the same order number for the 1958-63
              308 Win. ADL and 1969-80 6mm Rem. rifles.

25

Table 2
Carbines Produced by Caliber and Grade

Caliber	Grade	Years Made	Order Number	Production	Total by Caliber
.30-06	C	1960-63	5894	13,181	
	CDL	1961-63	5816	1,686	
	Standard	1964-67	9700	6,095	
	Standard	1968-80	9698	22,986	
					43,948
.308 Win.	C	1960-63	5906	2,588	
	CDL	1961-63	5908	518	
	Standard	1964-67	9702	1,955	
	Standard	1968-80	9714	8,930	
					13,991
.35 Rem.	Standard	1964-67	9696	456	
					456
.270 Win.	C	1960-63	5880	2,570	
	CDL	1961-63	5872	418	
					2,988
.280 Rem	C	1960-63	5886	1,222	
	CDL	1961-63	5914	121	
					1,343

Total carbine production   62,726

Table 3
Model 760 Cataloged Specifications

	Rifle	Carbine
Barrel length	22 inches	18-1/2 inches
Overall length	42 inches	38-1/2 inches
Weight	7-1/2 pounds	7-1/4 pounds*

Notes:
* The 1960 introduction gave the weight as 6-1/2 pounds, this was increased to 7 pounds in the 1961 catalog and 7-1/4 pounds in the 1962 catalog. The 1968 to 1976 catalogs gave the weight of the carbine as 6-3/4 pounds. It should be noted that there are normal variations in weight due to the density of wood, but these would be in the range of 1/4-pound.

Chapter 3
Model 740

The Model 740's introduction, originally scheduled for the same time as the Model 760, was delayed three years until December 1954. The public got its first view when the February 1, 1955 catalog was issued.

The Model 740 shared much of its development and design elements, such as the bolt, barrel extension, fire control and buttstock with the Model 760. This included the change from the original design to one based on the configuration of the 28 and 410 gauge Model 851/11-48 shotgun receiver. One of the reasons for the Model 740's delayed introduction was the design of a gas operating system. Three systems were tested in the late 1940s:

• Tappet — a piston in a cylinder moving one-eight to one-half inch and imparting momentum to the action bar as was accomplished in the US M1 Carbine used during the Second World War.

• Gas expansion — a piston in a cylinder moving a full stroke of about four inches as was used in the US M1 Garand rifle.

• Impulse reaction to gas flowing in a tube that impinges on a blind hole in the action bar.

Testing of the various gas systems was done in a caliber .30-06 Model 721 mule where each system was attached to the barrel. The systems did not

Model 740 A grade autoloader centerfire rifle. *Courtesy of Remington Arms Co.*

27

operate the bolt, but were attached to instruments measuring the force generated over time. The impulse reaction system was selected and by 1949 working models were being tested in calibers .30-06, .300 Savage and .270 Winchester. Much testing went into finding the proper size and length of the gas tube, barrel orifice and weight of the moving parts.

The new rifle, called Model 742 in-house, was officially changed on May 22, 1950 to the Model 740 Woodmaster. This appellation retained the name of the first successful autoloader, the Model 8, while the new model number denoted an entire new rifle.

By late 1950, prepilot testing of five Model 740s had revealed a need to revise the stock, especially the forearm, as well as the magazine , plus a need to hold the bolt open after the last shot and to release the bolt. Hard opening of the bolt had the engineers considering moving the operating handle from the bolt carrier to the action bar assembly.

The forearm and bolt handle continued to cause problems as further revisions were carried out and reported on April 16, 1953. The extractor also was being revised at this time.

The need for a coating to combat exhaust gas corrosion on the interior parts of the forearm was the subject of a February 26, 1954 report as these parts corroded with the use of commercial ammunition. The use of corrosive military surplus ammunition only hastened the process.

Field trials in 1951 noted

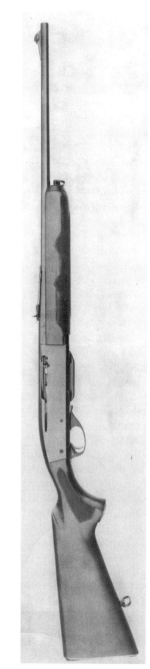

Model 740 ADL grade autoloader centerfire rifle. *Courtesy of Remington Arms Co.*

that the drop in the stock was too low for use with hunting scopes. It's interesting to note that these prepilot rifles were not drilled and tapped for mounting a scope; however, almost all of the testers had the rifles drilled and tapped for a scope.

L. Ray Crittendon, in 1951, developed the Model 740 magazine bolt release that generated much discussion. Testers liked that the magazine held the bolt open after the last round, but almost universally condemned the need to release the bolt in order to drop the empty magazine. Some preferred to use the Model 760 magazine, which allowed the bolt to close after the last shot. Remington's autoloader safety principle, used over the previous 40 years, won the day —to see the rifle open and empty after the last shot was fired. By 1953, 25 pilot Model 740s were issued for field testing by sales personnel. In general, comments were unanimously favorable with only the fore-end cap, stock and magazine generating any criticism. The new magazine with vertical ribs,

recently introduced for the Model 760, had its own problems. This magazine caused cartridges to jump out and jam the rifle. The cartridge shoulder wedged as it contacted the vertical rib forcing the cartridge over the feed lips so it would feed high in the chamber. It was decided on September 27, 1954 to use the older style, flat side magazine. The receiver stamping followed the pattern of the Model 760. *Remington* over the serial number was stamped in the lower middle of the left receiver panel while *Woodmaster* over Model 740 was stamped between the fire control pins. The two line barrel legend was stamped on the left side of the barrel above

Initial 1955 announcement flyer for the Model 740. *Author's collection.*

29

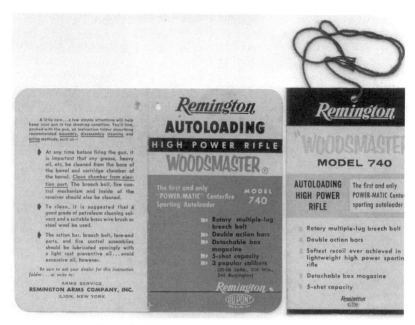

Two different Model 740 hang tags. *Courtesy of Roy Marcot.*

the fore-end:
REMINGTON ARMS CO. INC.,
ILION, N.Y. MADE IN U.S.A.
PATENT NO. 2,473,373
OTHERS PENDING

The caliber was stamped after the barrel legend.

1955

The first production Model 740 rifle, serial number 1001, was given to Remington Arms Company President C.K. Davis. It was an "F" grade with gold inlaid game scenes and is one of the rifles illustrated on the cover. The Model 740 was initially offered only in caliber .30-06 and in three grades:

• Grade A (standard)— $124.95
• Grade ADL (A Deluxe— $139.95
• Grade BDL (B Deluxe Special) — $159.95

The A grade had a low comb, plain, straight-grain walnut stock. Sling swivels were not furnished and the fore-end had a finger groove.

The ADL grade had a high comb, checkered, straight-grain walnut stock, grip cap, and sling swivels. The fore-end had a checkering pattern, called by today's collectors "5 Diamond" (not a factory designation), and the end cap bolt had a sling swivel.

The BDL grade was the same as the ADL with the addition of better wood. About 13 percent of the ADL and BDL rifles came with the low-comb stock option. The standard finish was lacquer.

Generally there were no markings on the early rifles to distinguish between the ADL and BDL grades. The original box end label had the grade noted on it and was the only

30

proof of an early BDL grade.

A few BDL grade rifles have been observed with the grade engraved after the model number. Many observed ADL grade rifles have nicely figured wood. As noted in the chapter on the Model 760, sometimes due to a flaw, BDL grade wood was downgraded and installed on ADL grade rifles.

The low comb stock had a 13-1/2 inch length of pull, 1-3/4 inch drop at the comb and 2-3/4 inch drop at the heel. The high comb stock had a 13-1/2 inch length of pull, 1-1/4 drop at the comb and 2 inch drop at the heel. The Model 740 had a 22 inch barrel with an overall length of 42-1/4 inches and weighed 7-1/2 pounds.

The front and rear sights were the same as those on Model 760, consisting of a ramp front sight with a white metal bead and a step-adjustable buckhorn rear sight dovetailed in the barrel. The difficulty of using the sights with a high comb stock was recognized early, and commented on, but no action was taken.

1956

The .308 Winchester cartridge chambering was announced May 1, with deliveries beginning in July.

The two line barrel legend stamped on the left side of the barrel above the fore-end had the patent number changed on September 26, 1956 to: REMINGTON ARMS CO. INC., ILION, N.Y. MADE IN U.S.A. PATENT NO. 2,685,754 OTHERS PENDING

1957

The .244 Remington cartridge chambering was

Remington's 1955 Big Game Rifles brochure advertising the autoloader Model 740 and pump action Model 760. *Author's collection.*

31

Instruction folder (above and next page) issued with each new Model 740 rifle. A single 12-1/2 inch by 13 inch page folded like a map to a 3-1/8 inch by 6-5/8 inch brochure. It is noteworthy that the instructions explicitly state this is a solid frame rifle and should not be dismounted except by a qualified gunsmith. A major omission in the instructions was the lack of any reference to fore-end float. It is essential that the fore-end float free of the receiver for accuracy. *Courtesy of Roy Marcot.*

announced in the January catalog.
1958
The .280 Remington cartridge chambering was available in the January catalog.

The BDL grade was dropped from the catalog.

A modified high comb stock, now called "all purpose," became standard in both the A and ADL grades. The all purpose stock had a 13-1/2 inch length of pull, 1-3/4 inch drop at the comb and a 2-1/4 inch drop at the heel.

The catalog continued to list low and high comb options. The 1958 and later production records separate the all purpose stocks from the low and high comb options. The ejection port cover was changed from steel to a black nylon resin called "Zytel."

The two line barrel legend stamped on the left side of the barrel above the fore-end was changed to reflect granting of additional patents:
REMINGTON ARMS CO. INC., ILION, N.Y. MADE IN U.S.A. PATENT No. 2,685,754 AND OTHERS
1959
The Model 740 was officially discontinued on December 31, 1959. However, according to a September 1 memorandum, around 1,000 Model 740 rifles in .244 Remington remained in the warehouse and sales continued through 1961 to clear the inventory. All caliber .244 Remington Model 740s had a 1-in-12" twist, which stabilized a 75 to 90 grain spitzer bullet. Customers wanted to use a 100

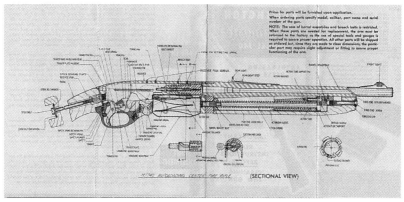

(SECTIONAL VIEW)

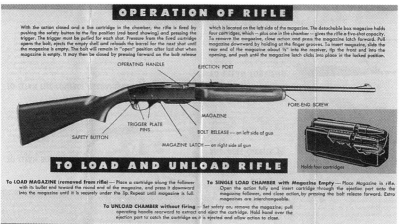

OPERATION OF RIFLE

With the action closed and a live cartridge in the chamber, the rifle is fired by pushing the safety button to the fire position (red band showing) and pressing the trigger. The trigger must be pulled for each shot. Pressure from the fired cartridge opens the bolt, ejects the empty shell and reloads the barrel for the next shot until the magazine is empty. The bolt will remain in "open" position after last shot when magazine is empty. It may then be closed by pressing forward on the bolt release

which is located on the left side of the magazine. The detachable box magazine holds four cartridges, which — plus one in the chamber — gives the rifle a five-shot capacity. To remove the magazine, close action and press the magazine latch forward. Pull magazine downward by holding ot the finger grooves. To insert magazine, slide the rear end of the magazine about ½" into the receiver, tip the front and into the opening, and push until the magazine latch clicks into place in the locked position.

OPERATING HANDLE

EJECTION PORT

FORE-END SCREW

MAGAZINE

TRIGGER PLATE PINS

BOLT RELEASE — on left side of gun

SAFETY BUTTON

MAGAZINE LATCH — on right side of gun

TO LOAD AND UNLOAD RIFLE

Holds four cartridges

To LOAD MAGAZINE (removed from rifle) — Place a cartridge along the follower with its bullet end toward the round end of the magazine, and press it downward into the magazine until it is securely under the lip. Repeat until magazine is full.

To SINGLE LOAD CHAMBER with Magazine Empty — Place Magazine in rifle. Open the action fully and insert cartridge through the ejection port onto the magazine follower, and close action by pressing the bolt release forward. Extra magazines are interchangeable.

To UNLOAD CHAMBER without firing — Set safety on; remove the magazine; pull operating handle rearward to extract and eject the cartridge. Hold hand over the ejection port to catch the cartridge as it is ejected and allow action to close.

grain spitzer bullet, which the 1-in-12" twist would not stabilize, so the caliber was not carried over to the new Model 742.

1960

The Improved Model 740, now called the Model 742, was introduced. The discontinued Model 740's serial number sequence began at 1001 and ended with 253821.

The total production of the Model 740, including 35 engraved rifles, was 252,275. Distribution by caliber was:

- .30-06 – 82 %
- .308 Win. – 11 %
- .280 Rem. – 5 %
- .244 Rem. – 2 %.

Table 1
Rifles Produced by Caliber and Grade

Caliber	Grade	Year Made	Production	Total by Caliber
.30-06	A-Lo	1955-57	123,646	
	ADL*	1955-57	41,417	
	BDL*	1955-57	671	
	A– all purpose	1958-59	30,744	
	ADL – all purpose	1958-59	11,028	
				207,506
.308 Win.	A-Lo	1956-57	14,711	
	ADL*	1956-57	3,746	
	BDL*	1956-57	57	
	A – all purpose	1958-59	6,901	
	ADL – all purpose	1958-59	1,840	
				27,255
.280 Rem	A-Lo	1957	8,832	
	ADL*	1957	3,160	
	BDL*	1957	32	
	A – all purpose	1958-59	127	
	ADL – all purpose	1958-59	9	
				12,160
.244 Rem.	A	1957	3,788	
	ADL*	1957	1,502	
	BDL*	1957	24	
	A – all purpose	1958	0	
	ADL – all purpose	1958	5	
				5,319
	D and F grade		35	
				35
Total rifle production				252,275

*The ADL and BDL figures include both high and low comb stocks.

Chapter 4

Model 742

The Model 742 began as a Model 740 improvement program in the spring of 1958. Persistent consumer complaints about cartridge ejection problems resulted in Remington conducting a study of the ejection path using high speed movies. The films showed that the bolt rebounded and struck the shell before the shell could clear the ejection port. In addition, the bolt caused the shell casing to be ejected toward the bottom of the ejection port. Remington's solution was to increase the distance between the ejector and the extractor, and reposition both the extractor and the ejector so that ejection occurred in the middle of the port. The width of the ejection port in the receiver also was increased.

Another problem, that of receiver rail upset, was corrected by fitting a latch between the bolt and the carrier to prevent bolt rotation when it was unlocked.

The second phase of the project, started in April 1959, increased the scope of the redesign to a new model, which was to be called the Model 742. Several specific areas of concern were addressed in the redesign. The first was improvement in the accuracy of the Model 742. Lack of accuracy and the inability to maintain zero and shoot consistently was a major complaint of consumers with the

Initial 1960 announcement flyer for the Model 742, front and back. *Courtesy of Jack Heath.*

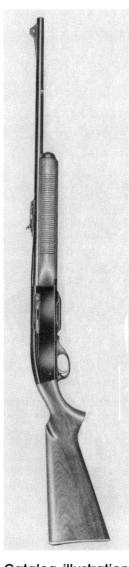

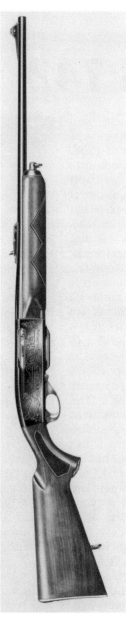

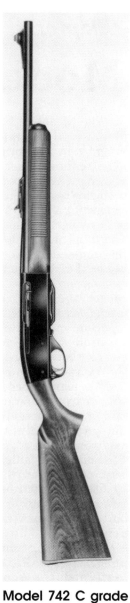

Catalog illustration of Model 742 A "Standard" grade rifle introduced in 1960. *Courtesy of Remington Arms Co.*

Model 742 ADL "Deluxe" grade rifle introduced in 1960. *Courtesy of Remington Arms Co.*

Model 742 C grade carbine. *Courtesy of Remington Arms Co.*

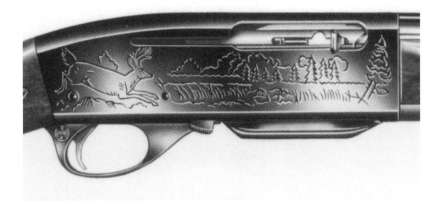

Close up of Model 742 ADL right receiver panel with deer scene introduced in 1960. Also found on Model 742 CDL grade carbines introduced in 1961. *Courtesy of Remington Arms Co.*

Model 740. Some of the problems were traced back to the fore-end adjustment which was not covered in the Model 740 manual. It was essential that the fore-end float free of the Model 740 receiver in order to assure the firearm's accuracy. The fore-end should be held away from the receiver and the screw tightened until the proper clearance of .020 inch was obtained.

The Model 742 used a double pitch fore-end screw that automatically made the adjustment. The screw was threaded 32 threads per inch in the receiver and 28 threads per inch in the fore-end bushing, allowing the fore-end to be pressed against the receiver, and as the screw was tightened, pulled forward to the final clearance of .026 inch. The Model 742 fore-end included a bushing with a spacer between the liner and the fore-end cap to prevent the fore-end bolt from pulling through and breaking the fore-end.

The second area of concern dealt with the improved ease of disassembly of the firearm. First, the operating handle pin was tipped from the vertical plane so that it could be driven out by a straight punch. The pin also was increased in diameter.

In addition, the breech ring nut was changed from one requiring a special spanner wrench to one with flats that allowed the use of a standard adjustable wrench. A nylon compression washer was fitted under the new nut to maintain tightness.

Lastly, the operating handle cross-section was changed from a square to cylindrical shape to reduce misalignment with the receiver slot and the resultant hard manual opening.

Third, a number of changes were made to the gas nozzle, rear sight and the magazine latch.

The fourth item to be redesigned was the magazine. It was necessary to raise the bullet points of the cartridges in the

Front of instruction folder issued with Model 742 carbine. Model 742 rifle is similar. *Courtesy of Jack Heath.*

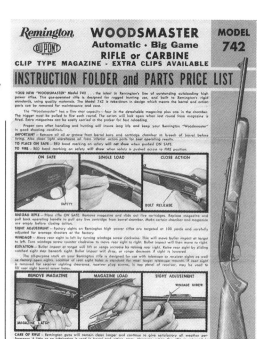

magazine so that they approached the centerline of the chamber. The magazine feed-lip configuration and dimensions were changed, and the magazines were identified by a boxed caliber designation.

Lastly, the Model 742 instructional folder was expanded to 22 photographs that illustrated proper maintenance, in contrast to the eight photos in the Model 740 manual.

Dick St. John, along with Charles Morse and Homer Young, were members of the Model 742 design team.

1960

The new Model 742, introduced on January 6, 1960 was offered in two grades. The Model 742 A was the standard grade with a plain, uncheckered buttstock and vertical grooves on

Detailed takedown instructions issued with each Model 742 rifle and carbine. *Courtesy of Jack Heath.*

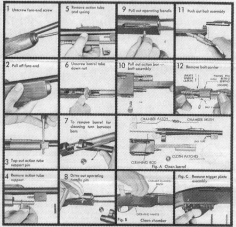

38

GREAT MOMENTS IN HUNTING

WITH A *Remington*,

MODEL 742

Stand up cardboard poster advertising the new "customized" Standard grade Model 742 rifle. The painting, from the Remington Arms Co. Collection, is by Robert Kuhn. *Author's collection.*

the fore-end. The Model 742 ADL "Deluxe" featured checkering on the buttstock and fore-end, sling swivels, a grip cap and a roll engraved game scene on both side panels.

The checkering on the fore-end was what today's collectors call the "5 Diamond" pattern (which is not a Remington designation). The shotgun style buttplate was aluminum with a bright finish on the edges. The catalog lists the stock specifications as:

- length of pull, 13-1/2 inches
- drop at comb, 1-1/4 inches
- drop at heel, 2 inches.

These were the same specifications as that of the previous Model 740"s high comb stock option. The few first year production Model 742 specimens examined by the author all had stocks measuring the same as the previous Model 740 "all purpose" stocks; that is, 13-1/2 inches length of pull, 1-1/4 inches drop at the comb and 2-1/4 inches drop at the heel. Later catalogs changed the specifications to the "all purpose" stocks.

The Model 742 A cost $138.50 and the Model 742 ADL, $154.45.

The stamping of the Model 742 A left receiver panel followed the pattern of the Models 740 and 760 with *Remington* stamped over the serial number in the middle while *Woodmaster* over Model 742 was stamped between the fire control pins.

The Model 742 ADL "Deluxe" had roll marked scenes of a deer on the right receiver panel and a bear on the left panel. The left panel had *Remington* over *Woodmaster* over Model 742 over the serial number stamped between the fire control retaining pins.

The two line barrel legend was stamped on the left side of the barrel above the fore-end.
REMINGTON ARMS CO. INC., ILION, NY MADE IN U.S.A.
PATENT No. 2,685,754 AND OTHERS

The caliber was stamped after the barrel legend. The Models 742 A and 742 ADL were initially offered in .30-06, .308 Winchester and .280 Remington.

A new screw-attached rear sight was step adjustable for elevation, screw adjustable for windage, and set on a base to have a useable line of sight with the "all purpose" stock. The ramp front sight was now a flat-faced gold bead.

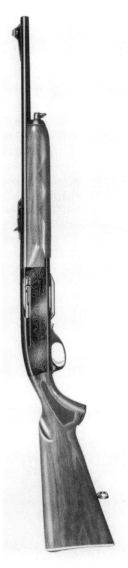

Model 742 CDL Deluxe grade carbine introduced in 1961. *Courtesy of Remington Arms Co.*

Model 742 "customized" Standard grade rifle with pressed fleur-de-lis checkering, introduced in 1964. *Courtesy of Remington Arms Co.*

Model 742 BDL Custom Deluxe rifle with stepped receiver, Monte Carlo cheekpiece buttstock, flat fore-end and basket weave checkering pattern. Introduced in 1966. *Courtesy of Remington Arms Co.*

1966 announcement of Models 742 and 760 BDL grade rifles. *Author's collection.*

1961

Model 742 C and Model 742 CDL carbines were introduced in the same calibers and with the same features as the rifles. The CDL grade carbine had the same game scenes and markings as the ADL grade rifles. The catalog description noted that the carbine had an 18-1/2 inch barrel, an overall length of 38-1/2 inches, and weighed 7 pounds. 742 CARBINE was stamped in large letters after the caliber designation on the barrel.

The price for A grade rifles and C grade carbines was increased to $139.95, while the ADL grade rifles and CDL grade carbines were increased to $155.95.

1962

The Model 742 carbine weight was increased to 7-1/4 pounds, according to the Remington catalog. All early Model 742 carbines examined by the author weighed about 7-1/4 pounds. It should be noted that normal variations in wood density result in weight changes in the range of about one quarter of a pound.

1963

Caliber 6mm Remington was added to the rifle line. A few early rifles were stamped 6MM MAG with MAG exed over. The 6mm Remington cartridge case had the same specifications as the .244 Remington case but offered the desired 100 grain spitzer bullet. The twist was increased to 1 in 9-1/4 inches to handle the longer bullet.

The ADL and CDL receiver panel game scenes were discontinued at the end of the year.

1964

Caliber .280 Remington was dropped from the list of calibers available in the carbine line.

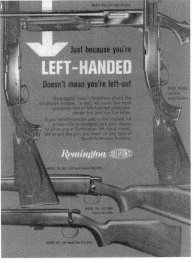

1966 announcement of left hand buttstock option for the Model 742 and 760 BDL grade rifles. *Author's collection.*

1966 150th Anniversary standup cardboard poster advertising the standard grade Model 742 rifle. The painting ˵Bull of the Woods˝, from the Remington Arms Co. Collection, is by Robert Kuhn. *Author´s collection.*

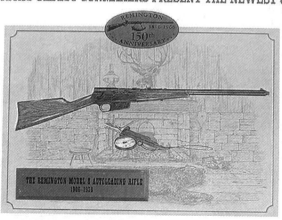

1966 150th Anniversary standup cardboard poster advertising the BDL grade Model 742 rifle. The painting of the Model 8 Autoloading Rifle is by James K. Laing. *Author´s collection.*

A single "customized" Standard grade rifle and carbine with newly-designed pressed checkering patterns on the buttstock and fore-end replaced the A and ADL grade rifles and the C and CDL grade carbines.

1965

No changes were noted.

1966

The Model 742 BDL Deluxe grade was offered in calibers .30-06 and .308 Winchester. The rear of the Model 742 BDL receiver had a distinctive step and the Monte Carlo buttstock, with a choice of right hand or left hand cheek piece, as well as the flat fore-end, had a basket-weave checkering pattern.

The left hand models had the safety reversed; however, ejection was still to the right side. The BDL grade stocks have a 13-5/16 inch length of pull, 1-5/8 inch drop at comb and a 2-1/2 inch drop at heel.

Also, a 150th Anniversary Model, in caliber .30-06, was offered. See the chapter on Commemorative Rifles for more details on this model.

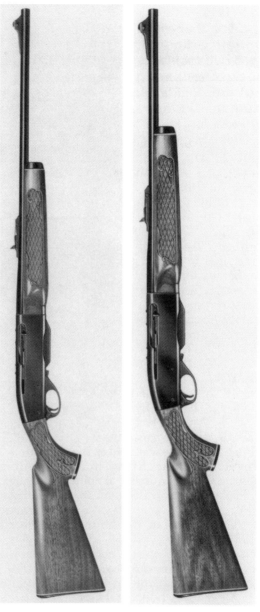

1968 restyled Standard grade Model 742 rifle, left, and Carbine, right, both with skip-line pressed checkering pattern. *Courtesy of Remington Arms Co.*

43

Top: 1973 buttstock skip-line checkering pattern. Bottom: 1968 buttstock skip-line checkering pattern. *Photo: Author. Author's collection.*

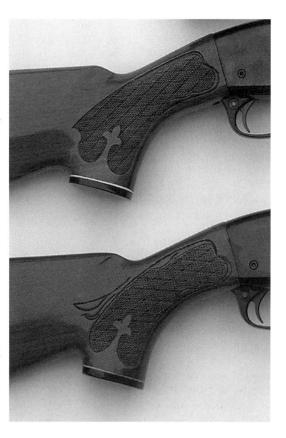

All models now had a Du Pont-developed RK-W gloss wood finish.

A Standard grade Model 742 rifle, serial number 312376, caliber .30-06, was presented to John T. Amber, editor of *Gun Digest.* A large presentation banner was engraved on the left receiver panel — "PRESENTED TO / JOHN AMBER / BY REMINGTON ARMS COMPANY / AT THE 1967 / FIREARMS EDITORS' SEMINAR / Merrymeeting Bay, Maine". The rifle was sold at the November 1986 Richard A. Bourne Co. Inc. auction of John T. Amber's arms collection.

1967

The Canadian Centennial Model 742 in caliber .308 Winchester was offered. See the chapter on Commemorative Rifles for more details.

All models had the barrel legend on the left side of the barrel above the fore-end changed to three lines to include additional patents:

REMINGTON ARMS CO. INC., ILION, N.Y. MADE IN U.S.A.

PATENT NO. 2,685,754 - 2,585,195—2,473,313– 2,675,638 – 2,473,313 AND OTHERS

1968

All models had metal parts finished by vibra-honing to make the interior metal surfaces mirror smooth. The Standard grade rifle and carbine had a restyled buttstock and forearm with pressed checkered in a new skip-line pattern. The aluminum buttplate was changed to black plastic and it, as well as the metal fore-end cap and black plastic grip cap had white line spacers.

The stock had a 13-5/16 inches length of pull, 1-5/8 inches drop at comb and 2-1/4 inches drop at heel.

Caliber .243 Winchester

44

was added to the list of rifle calibers available.

The Model 742 serial number sequence was changed on November 26, 1968 as a result of the 1968 Gun Control Act, which required that no two firearms from the same manufacturer have the same serial number. The initial Model 742 serial number sequence began at 1001 and ended at 396562. The new serial number sequence was shared with the Model 760 pump action rifle and began at A6900000.

1969

The Model 742 BDL grade rifle was changed to "Custom Deluxe".

1970

No changes were noted.

1971

All models had internal parts coated with Teflon "S" to reduce friction and dirt buildup.

Eight Model 742s were tested in caliber .25-06 Remington, but the cartridge was not added to the list of rifle calibers available.

1972

No changes were noted.

1973

There was a minor change in the Standard grade checkering pattern. The Standard grade stock dimensions were listed in the catalog as 13-1/4 inches, length of pull; 1-5/8 inches, drop at the comb; and 2-1/4 inches, drop at the heel.

1974

The barrel legend stamped on the left side of the barrel above the fore-end was changed back to two lines:
REMINGTON ARMS CO. INC., ILION, N.Y.
MADE IN U.S.A. PAT. NUMBER – 2,473,373

1975

The rear sight for all models was changed to a sliding ramp style.

1976

A Bicentennial Model 742, in caliber .30-06, was featured in a special brochure. See the chapter on Commemorative Rifles for more details.

The 1 millionth Model 742, manufactured during December 1976, was issued a special serial number – A1000000 — and was engraved as an F grade with gold inlaid game scenes. It currently is exhibited in the Remington Arms Company Museum in Ilion, NY.

1977

No changes were noted.

1978

No changes were noted.

1979

No changes were noted.

Remington Arms Co. tie clasp, 3 inches long in the shape of a Model 742 BDL rifle. It was also available in a smaller 2-1/4 inch version. *Author's collection.*

1980

Caliber .280 Remington, originally introduced in 1960, was renamed 7mm Express Remington. However, this would last only until 1983 when it reverted to the original caliber .280 Remington designation

The Model 742 was discontinued in all grades and calibers December 31, 1980 with the introduction of the Model Four and Model 7400 autoloader rifles.

1981

Sales of the Model 742 continued well into the year to clean out warehouse stock.

The total production of the Model 742, including rifles, carbines, commemoratives and 75 engraved arms, was 1,497,169. Distribution by caliber was:

- .30-06 – 75%
- .308 Win. – 12%
- .243 Win. – 6%
- .280 Rem./7 mm Exp. Rem. – 5%
- 6 mm Rem. – 2%.

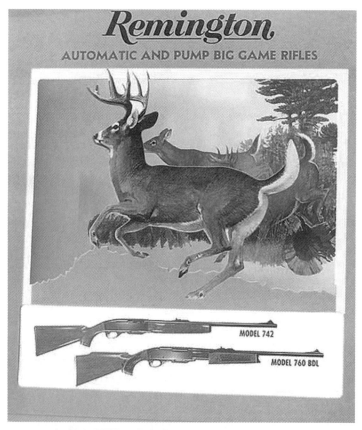

An unusual mid-1970s standup, three-dimensional cardboard poster advertising the Model 742 and 760 rifles. The curved background painting is by Robert Kuhn. The jumping buck in the foreground is separate. *Author's collection.*

Table 1
Rifles Produced by Caliber and Grade

Caliber	Grade	Year Made	Order Number	Production	Total by Caliber
.30-06	A	1960-63	5838	52,774	
	ADL	1960-63	5840	22,952	
	Standard	1964-67	9654	106,162	
	150th Ann.	1966		11,412	
	BDL*	1966-80	9666/9668	163,899	
	Standard	1968-80	9636	676,899	
	Bicentennial	1976	7612	10,108	
					1,044,206
.308 Win.	A	1960-63	5846	13,495	
	ADL	1960-63	5848	4,729	
	Standard	1964-67	9656	20,057	
	BDL*	1966-80	9640/9644	24,254	
	Canadian Cen	1967		1,968	
	Standard	1968-80	9638	88,665	
					153,168
.280 Rem.	A	1960-63	5830	3,621	
	ADL	1960-63	5832	1,715	
	Standard	1964-67	9652	3,486	
	Standard**	1968-79	9634	53,325	
	Standard***	1980	9630	17,504	
					79,651
6mm Rem.	A	1963	5866	1,308	
	ADL	1963	5868	855	
	Standard	1964-67	9650	3,651	
	Standard	1968-80	9632	18,499	
					24,313
.243 Win.	Standard	1968-80	9648	81,786	
					81,786
	D and F grades	1960-80		75	
					75
Total rifle production					1,383,199

Notes: *Includes right and left handed rifles.
 **Includes order #9635 – European market special order with 2 shot magazine.
 ***Caliber designation changed to 7 mm Express Remington.
Includes order #9631 – European market special order with 2-shot magazine.

Table 2
Carbines Produced by Caliber and Grade

Caliber	Grade	Year Made	Order Number	Production	Total by Caliber
.30-06	C	1961-63	5858	10,945	
	CDL	1961-63	5860	4,112	
	Standard	1964-67	9662	15,828	
	Standard	1968-80	9670	54,176	
					85,061
.308 Win.	C	1961-63	5862	2,801	
	CDL	1961-63	5864	1,070	
	Standard	1964-67	9664	5,824	
	Standard	1968-80	9672	18,552	
					28,247
.280 Rem.	C	1961-63	5854	476	
	CDL	1961-63	5856	186	
					662

Total carbine production   113,970

Table 3
Model 742 Cataloged Specifications

	Rifle	Carbine
Barrel length	22 inches	18-1/2 inches
Overall length	42 inches	38-1/2 inches
Weight	7-1/2 pounds	7-1/4 pounds*

Notes:
*The 1968 – 1976 catalogs have a weight of 6-3/4 pounds for the carbine. The author has no explanation for the variation in weight. So far all Model 742 carbines examined have weighed just over 7 pounds. It should be noted that there are normal variations in weight due to the density of wood, but these would be in the range of a quarter of a pound.

Chapter 5 —
New Generation Rifles

Remington, in 1974, began a major product improvement program to replace the Model 742 and Model 760. By 1976 development of the "New Generation Rifles," as the new autoloader and pump action rifles were called, was well under way. The new rifles — Models Four, Six, 7400 and 7600 — were introduced in late 1980. The initial 1981 advertising, both in the catalog and in the sporting press, concentrated on the deluxe grade autoloader Model Four and pump action Model Six. It wasn't until the 1983 catalog that the popular-priced autoloader Model 7400 and pump action Model 7600 received equal billing.

In mid-1984, a new series of budget-priced rifles and shotguns, directed toward the mass merchandisers, was introduced with its own brochure. The line included the Sportsman 74 autoloader and Sportsman 76 pump action rifles. The difference between the various models was in the level of metal finish and wood.

Remington's

Bolt assembly illustrated in the 1982 catalog. *Author's collection.*

December 1, 1980 writer's seminar previewed the new rifles that were to be introduced in 1981, and the following improvements were noted in the official news release:

One of the major innovations in both of Remington's new rifles, the Model Four autoloader and the Model Six pump action, involves a totally new locking system design.

In comparison with their predecessors, the Model Four and the Model Six feature a simpler, more rugged bolt assembly. The number of locking lugs on the breech bolt has been reduced from nineteen to four. The existence of fewer and larger locking lugs on the bolt provides a number of advantages. First, the simplified design permits better dimensional control to desired tolerances during manufacture. This leads not only to more positive locking of the action, but to much smoother locking and unlocking operation.

A bolt you'd expect in a bolt action.

A new, hardened steel receiver insert has a smooth, recessed groove that mates with a matching lug on top of the bolt carrier. Combined with top and bottom support of deeper and heavier action bars by channels in the side of the receiver, this arrangement provides a highly stable, three-point guidance system for the integral bolt carrier/action bar unit. The increased stability that results from this three-point support and guidance adds to smoother operation of the entire action and increases substantially the durability and longevity of the rifles.

Another improvement designed to further extend strength and durability of both rifles is the combining of a two-piece breech ring and barrel extension into a heavier, single piece, milled from a solid piece of steel.

Because of the potential for more rapid operation of both autoloading and pump actions, smooth, properly directed cartridge feeding is essential. To help ensure trouble-free cartridge guidance into the chamber, both rifles include a funnel-shaped, 360-degree counterbore at the rear of the barrel to provide a sloping entrance ramp for cartridge feeding. A newly designed magazine box of heavier and

New, larger magazine release illustrated in the 1982 catalog. Author's collection.

stronger metal also aids in feeding cartridges at a shallower angle. Improved dimensions of the magazine release now make it easier to operate, even with gloves.

As an additional new feature common to both Model Four and Model Six rifles, the bolts are now fitted with Remington's new rivetless extractor that eliminates the need for a rivet and a rivet cut through the bolt lip. This improved design adds to bolt strength and cartridge support, extends extractor life considerably and makes extractor replacement much easier if necessary.

In the new Model Four autoloading rifles, several design changes create a significant improvement in gas sealing and overall performance. In place of a previous design in which high-pressure gas for autoloading operation was sealed by multiple-machined flat surfaces, the Model Four now utilizes a self-centering conical gas seal between the barrel lug nozzle and the inertia sleeve. The result is improved gas sealing and more consistent bolt velocity.

The bolt improvements and reduction of binding between the bolt carrier assembly and

Our smoothest magazine release yet.

receiver eliminated the need for the bolt latch assembly. The feeding malfunction called "stem chamber," which occurs when the bullet of the feeding cartridge is stopped by any of the surfaces at the rear of the chamber, was significantly reduced by the funnel-shaped 360 degree counterbore, new magazine and moving the chamber forward one quarter of an inch.

The new magazines had the caliber designation in a box with rounded ends.

Both the screw size and screw hole spacing for the scope mounts were changed and neither was interchangeable with previous models. The hole spacing was increased from 3-1/8 inches to 3-9/16 inches. The screw size for the scope mounts, as well as for the front and rear sights, was increased to 8-40. The front and rear sights use the same screw size and the receiver plug screws can be used to fill the tapped holes in the barrel when the sights are removed.

It was recognized in 1984 that fielding three grades, each with different names, of both the autoloader and pump action New Generation Rifles was not the best marketing strategy, so it was proposed that the line be collapsed within two or three years. The Model Four, Model Six, Sportsman 74 and Sportsman 76 were dropped from the 1988 catalog.

The design team consisted of Section Manager Jim Martin, Merle Carter, Albert Eddy, Jim Hutton and Jack Kost.

Chapter 6 — **Models 7400 and 7600**

The popular-priced autoloader Model 7400 and pump action Model 7600, both introduced in January of 1981, are still in production. They replaced the standard grade Model 742 autoloader and Model 760 pump action rifles. The Model 7400 and Model 7600 outsold the higher-priced Model Four and Model Six by a wide margin. The Model 7400's initial retail price was $399.95 and the Model 7600's was $349.95.

The Model 7400 and Model 7600 differed from the Models Four and Six rifles only in cosmetic details. They retained the Model 742 and Model 760 straight comb buttstock shape and added fine line press checkering in a fleur-de-lis pattern. The Model 7600 fore-end was the same shape as the Model 760, but with the new checkering pattern.

The Model 7400 fore-end was reshaped with a flared bottom for better grip and checkered in the new pattern. The initial stock specifications were:

• Model 7400 — length of pull, 13-3/8 inches; drop at comb, 1-13/16 inches; and drop at heel 2-1/4 inches

• Model 7600 — length of pull, 13-3/8 inches; drop at comb, 1-11/16 inches; and drop at heel, 2-1/8 inches.

The 1987 to 1989 catalogs noted different stock specifications. The author has no explanations for these variations, as all examined specimens of the Model 7400 and 7600 have the same stock specifications as those noted by Remington for its Model 7400.

Remington over the serial number was stamped on the left receiver panel, while *Model 7400* or *Model 7600* was stamped between the fire control pins. The trade names, *Gamemaster*

Fine-line engraving applied to left receiver panel of Model 7600s beginning in 1996. The Model 7400 is similar with a pair of elk replacing the bears. *Photo: Author. Author's collection.*

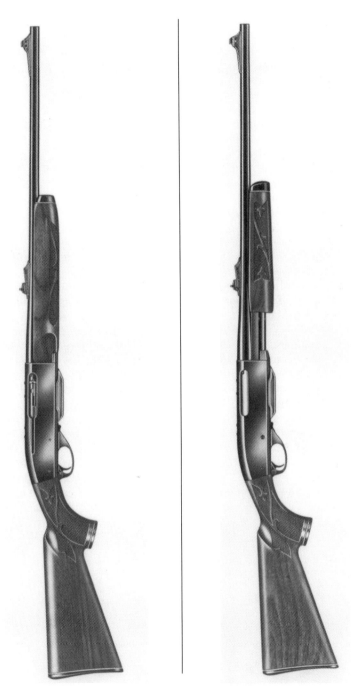

Model 7400 (left) and Model 7600 (right) rifles with straight comb stocks, introduced in 1981. *Courtesy of Remington Arms Company.*

and *Woodsmaster,* were not continued with these models. The two line barrel legend was stamped on the left side of the barrel above the fore-end. The caliber was stamped after the barrel legend.

Both the Model 7400 and the Model 7600 are current models and production data is not available. In some cases estimates have been made based on the author's database and discussions with dealers and distributors.

1981

The Model 7400 was initially offered in the following calibers – .30-06, .308 Winchester, 7mm Express Remington, .270 Winchester, .243 Winchester and 6mm Remington. All calibers except the 7mm Express Remington also were chambered in the Model 7600.

The two line barrel legend was stamped on the left side of the barrel above the forearm: REMINGTON ARMS CO. INC. ILION, N.Y. MADE IN U.S.A.

1982

No changes were noted.

1983

The 7mm Express Remington cartridge returned to its original name — .280 Remington.

1984

The two line barrel legend was changed to:
WARNING – READ INSTRUCTION BOOK FOR SAFE OPERATION – FREE FROM *****REMINGTON ARMS COMPANY, INC., ILION, N.Y., U.S*****

1985

No changes were noted.

1986

No changes were noted.

1987

The Model 7600 carbine in caliber .30-06 was introduced. The 7-1/4 pound carbine had an overall length of 38-1/2 inches and a barrel length of 18-1/2 inches. *7600 CARBINE* was stamped in large letters after the caliber designation on the barrel.

The 6mm Remington was dropped from the Model 7600's list of available cartridges.

1988

The Model 7400 carbine in caliber .30-06 was introduced. The 7-1/4 pound carbine had an overall length of 38-1/2 inches and a barrel length of 18-1/2

Fine-line engraving applied to right receiver panel of Model 7600s beginning in 1996. The Model 7400 is similar with a pair of rams replacing the deer. *Photo: Author. Author's collection.*

54

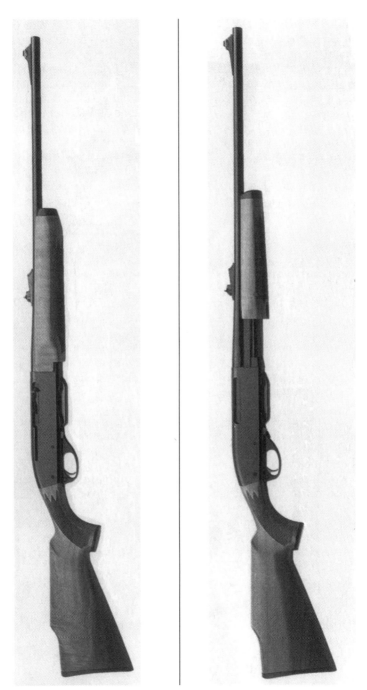

Model 7400 (left) and Model 7600 (right) rifles with Monte Carlo style stocks and cut checkering. Introduced in 1991. *Courtesy of Remington Arms Co.*

inches. *7400 CARBINE* was stamped in large letters after the caliber designation on the barrel.

The Model 7600 added the .280 Remington and the .35 Whelen to the list of available cartridges.

The Model 7400 dropped the 6mm Remington from the list of available cartridges.

1989

The 1989 catalog noted that the standard stock finish for both models was now satin, with no change in order number.

1990

The overall length of all models was increased by 5/8 of an inch to 42-5/8 inches for rifles and 39-1/8 inches for carbines.

Distributors Grice Wholesale and R. M. Crawford each special ordered 500 non-cataloged Remington Model 7600 pump action rifles in caliber 7mm-08 Remington.

1991

The first major change in the Model 7400 autoloader and Model 7600 pump action rifles and carbines was an upgrade in the stocks and checkering. The restyled stocks featured a Monte Carlo style buttstock with a new pattern of cut checkering. Standard wood finish was satin; however, rifles in calibers .30-06 and .270 Winchester were available with a gloss wood finish.

The stock specifications, still in use today, are the same as the original Model 7400 — length of pull, 13-3/8 inches; drop at comb, 1-13/16 inches; and drop at heel 2-1/4 inches.

Calibers offered in both models were .30-06, .308 Winchester .280 Remington, .270 Winchester and .243 Winchester, plus the .35 Whelen in the Model 7600. The suggested retail price was $501.00 for the Model 7400 and $484.00 for the Model 7600.

A non-cataloged Model 7600, in calibers .25-06 Remington and .257 Roberts, was special ordered by Grice Wholesale.

A gloss stock finish caliber .30-06 Model 7400 with the 175th Anniversary logo roll-marked on the left receiver panel was cataloged. A non-cataloged 175th Anniversary Model 7400, chambered for the .270 Winchester cartridge, was special ordered by a distributor. Grice Wholesale special ordered a 175th Anniversary Model 7600 chambered for the 7mm-08 Remington cartridge. See the chapter on Commemorative Rifles for more details on the 175th Anniversary Model.

1992

Grice Wholesale offered the non-cataloged "Deer Hunter" Model 7600 pump action rifle in calibers .30-06, 7mm-08 Remington and 270 Winchester. The B/O Special "Deer Hunter" had a gold-filled roll-marked deer scene on the left receiver panel. See the chapter on Commemorative rifles for details.

Grice Wholesale continued to offer non-cataloged Model 7600s in calibers .25-06 Remington and .257 Roberts.

1993

The next series of changes occurred in 1993 with the introduction of a new series of rifles. The Model 7400 SP and Model 7600 SP or "Special Purpose" rifles featured matte finished metal, a low-luster wood finish, sling swivels and a camouflage Cordura sling. The SP stocks have the same configuration as the regular Model 7400 and Model 7600 and both SP versions were offered in calibers .30-06 and .270 Winchester.

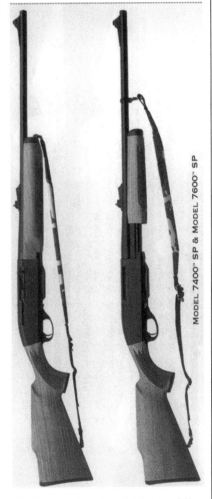

MODEL 7400" SP & MODEL 7600" SP

1994 catalog illustration of the Models 7400 and 7600 Special Purpose Rifles. These rifles have matte finished metal and low luster wood. *Courtesy of Remington Arms Company.*

An airline approved hard case was included with the purchase price of the satin or gloss finished Models 7400 and 7600.

The .35 Whelen was added to the Model 7400 list of available calibers.

Grice Wholesale continued to offer non-cataloged Model 7600s in calibers .25-06 Remington and .257 Roberts.

1994

Non-cataloged, black laminated stocked Model 7600 pump action rifles in calibers .30-06, .270 Winchester and .243 Winchester were offered by Grice Wholesale. A non-cataloged high gloss Model 7600 in caliber .243 Winchester was offered by another distributor.

1995

The catalog dropped the Special Purpose Models 7400 SP and 7600 SP.

Grice Wholesale continued to offer non-cataloged black laminated stocked Model 7600s in calibers .30-06, .270 Winchester and .243 Winchester.

1996

The receivers of the Model 7400 and Model 7600 were embellished with pressure applied fine-line rolled engraving. This was noted on the box end label as "new enhanced receiver engraving," later changed to "enhanced receiver engraving."

The Model 7400 receiver featured a pair of elk on the left panel and a pair of rams on the right panel. The Model 7600 had two deer on the right panel and two bears on the left panel.

The Model 7400 dropped the .35 Whelen cartridge from the list of available cartridges.

57

Models 7400 and 7600's with synthetic stocks introduced in 1998.
Courtesy of Remington Arms Company.

1997

The "Buckmasters American Deer Foundation" Limited Edition Model 7400 and Model 7600 were featured in the 1992 catalog. See the chapter on Commemorative Rifles for further details.

The regular Model 7400 cost $573 and the Model 7600 $540.

The Model 7600 dropped the .35 Whelen cartridge from the list of available cartridges.

1998

Synthetic stocked rifles were introduced in the 1998 catalog. The fiberglass-reinforced buttstock and forearm, as well as the metal work, had a matte black non-reflective finish. The same calibers were offered as in the regular line. The synthetic stocked Model 7400 cost $473 and the synthetic stocked Model 7600, $440.

Model 7600s in non-cataloged calibers 7mm-08 Remington, .260 Remington, .25-06 Remington and 6mm Remington, were offered by Grice Wholesale. These had the enhanced receivers with the pressure applied fine-line engraving.

1999

The "Integrated Security System" (ISS) trigger lock, manufactured as a integral component of the Model 7400 and Model 7600 safety, was introduced in late 1999.

2000

A new magazine with a separate matte black plastic bottom was introduced in late 2000.

The barrel legend was changed to a single line:
REMINGTON ARMS COMPANY INC. ILION, NY.

Non-cataloged brown laminated stocked, matte blue finished metal Model 7600 pump action rifles in calibers .30-06, 7mm-08 Remington, .270 Winchester and .243 Winchester were offered by Grice Wholesale.

2001

The Model 7400 and Model 7600 discontinued the .280 Remington chambering.

Non-cataloged synthetic stocked, matte finished metal Model 7600 pump action rifles and carbines in caliber .35 Remington were offered by Grice Wholesale.

Integrated Security System (ISS) trigger lock in the locked position, top, and with the key inserted, bottom. Introduced in late 1999 on Models 7400 and 7600. *Photos: Author. Courtesy of S. Scribner.*

B/O SPECIALS and NON-CATALOGED MODELS

Remington Arms Co. applies a red B/O SPECIAL label on the shipping box end label to identify back-ordered special order rifles and carbines such as the 1992 Grice "Deer Hunter" Model 7600 rifle and the 1999-2000 Benoit Commemorative Model 7600 carbine. See the chapter on Commemorative Rifles for a picture of the Benoit Commemorative carbine shipping box end label.

The onset of "niche" marketing in the early 1990s resulted in a wider selection of rifles tailored to specific hunter's needs or desires. Many new variations, not cataloged, were special-ordered by the various distributors in quantities ranging from 75 to around 1,200 rifles. The majority of collectors are not aware of these limited production variations

The author, over the past four years, has noticed several non-cataloged variations of Model 7400 and 7600 rifles for sale in trade papers such as Shotgun News and Gun List, as well as on a number of internet sites. These included:

Model 7400
• High Gloss .243 Winchester
• Special Purpose carbine .30-06
• Synthetic carbine .280 Remington

Model 7600
• High Gloss .243 Winchester

Production data is not available on these non-cataloged variations.

Grice Wholesale Special Order Model 7600

Grice Wholesale, Clearfield, Pennsylvania has offered the greatest number of these non-cataloged variations over the years. They include:

Year(s)	Order Number	Caliber	Quantity	Notes
1990	4668	7mm-08 Rem.	500	
1991	4672	7mm-08 Rem.	1,000	175th Anniversary logo
1991/93	4674	.25-06 Rem.	1,200	
1991/93	4676	.257 Robts.	1,200	
1992	4666	.270 Win.	500	Deer Hunter Special
1992	4668	7mm-08 Rem.	1,000	Deer Hunter Special
1992	4670	.30-06	500	Deer Hunter Special
1994/95	5138	.243 Win.	75	Black laminated stocks
1994/95	5139	.270 Win.	700	Black laminated stocks
1994/95	5141	.30-06	700	Black laminated stocks
1998	4650	6mm Rem.	250	Satin walnut stocks, enhanced receiver
1998	5137	.260 Rem.	250	Satin walnut stocks, enhanced receiver
1998	4674	.25-06 Rem.	250	Satin walnut stocks, enhanced receiver
1998	4668	7mm-08 Rem.	250	Satin walnut stocks, enhanced receiver
2000	5163	.243 Win.	250	Brown laminated stocks, matte blue finished metal
2000	5165	.270 Win.	250	Brown laminated stocks, matte blue finished metal
2000	5167	.30-06	250	Brown laminated stocks, matte blue finished metal
2000	5169	7mm-08 Rem.	250	Brown laminated stocks, matte blue finished metal
2001	N/A	.35 Rem	300	Rifle, synthetic stocks, matte finished metal
2001	N/A	.35 Rem.	200	Carbine, synthetic stocks, matte finished metal

Table 1
1981-1990 Cataloged Models

Model	Caliber	Order Number	Date Introduced	Date Discontinued
7400	.30-06	4716	1981	1991
7400	.308 Win.	4718	1981	1991
7400	7mm Exp Rem. /.280 Rem.	4714	1981	1991
7400	.270 Win.	4712	1981	1991
7400	.243 Win.	4710	1981	1991
7400 Carbine	.30-06	4702	1988	1991
7600	.30-06	4656	1981	1991
7600	.308 Win.	4658	1981	1991
7600	.270 Win.	4654	1981	1991
7600	.243 Win.	4652	1981	1991
7600	6mm Rem.	4650	1981	1987
7600 Carbine	.30-06	4660	1987	1991
7600	.280 Rem.	4662	1988	1991
7600	.35 Whelen	4664	1988	1991

Note: Stock finish changed to satin in 1989 with no change in order number.

Table 2
Model 7400 and Model 7600 Catalog Specifications

	Rifle	Carbine
1981-1989		
Barrel length	22 inches	18-1/2 inches
Overall length	42 inches	38-1/2 inches
Weight	7-1/2 pounds	7-1/4 pounds
1990 – present		
Barrel length	22 inches	18-1/2 inches
Overall length	42-5/8 inches	39-1/8 inches
Weight	7-1/2 pounds	7-1/4 pounds

Note: The 1988 and 1989 catalogs reported a weight of 8 pounds for the Model 7600 chambered in .35 Whelen. The 1990 catalog dropped that weight to 7 pounds. The 1991 and later catalogs report the weight for the caliber .35 Whelen Model 7600 as 7-1/4 pounds. The few Model 7600 pump action rifles in .35 Whelen examined by the author all seem to be around 7-1/4 pounds. It should be noted that there are normal variations in weight due to the density of wood, but these would be in the range of a quarter of a pound.

Table 3
1991 – 2001 Cataloged Models

Model	Caliber	Order Number Satin	Gloss	Date Introduced	Date Discontinued
7400	.30-06	9763	9765	1991	
7400	.308 Win.	9767		1991	
7400	.280 Rem	9761		1991	2001
7400	.270 Win.	9757	9759	1991	
7400	.243 Win.	9755		1991	
7400 Carbine	.30-06	9769		1991	
7400 175th	.30-06	4700		1991	1992
7400 SP	.30-06	4706		1993	1995
7400 SP	.270 Win.	4704		1993	1995
7400	.35 Whelen	4699		1993	1996
7400 ADF	.30-06	4695		1997	1998
7400 Syn	.30-06	9796		1998	
7400 Syn	.308 Win.	9797		1998	
7400 Syn	.280 Rem	9795		1998	2001
7400 Syn	.270 Win..	9794		1998	
7400 Syn	.243 Win.	9793		1998	
7400 Syn Carb	.30-06	9783		1998	
7600	.30-06	4657	4671	1991	
7600	.308 Win.	4659		1991	
7600	.280 Rem.	4663		1991	2001
7600	.270 Win.	4655	4667	1991	
7600	.243 Win.	4653		1991	
7600 Carbine	.30-06	4661		1991	
7600	.35 Whelen	4665		1991	1997
7600 SP	.30-06	4679		1993	1995
7600 SP	.270 Win.	4677		1993	1995
7600 ADF	.30-06	4669		1997	1998
7600 Syn	.30-06	5149		1998	
7600 Syn	.308 Win.	5151		1998	
7600 Syn	.280 Rem	5147		1998	2001
7600 Syn	.270 Win.	5145		1998	
7600 Syn	.243 Win.	5143		1998	
7600 Syn Carb	.30-06	5153		1998	

European Market Specifications:
Model 7400 and Model 7600 rifles and carbines, with 2 shot magazines, have been produced in calibers .243 Winchester, .270 Winchester, 7mm Express Remington/ .280 Remington, .30-06 and .35 Whelen for the European market. They have been made with wood or synthetic stocks and with plain or Monte Carlo style buttstocks. The author has copies of a French language Remington catalog and parts manual listing Rivolier SA, BP42- 42170 ST-JUST-RAMBERT-FRANCE as their agent.

Chapter 7 — **Model Four and Model Six**

The autoloader Model Four and pump action Model Six centerfire rifles were introduced in January 1981 as replacements for the Models 742 and 760 BDL Custom Deluxe rifles. Remington's initial advertising, both in the 1981 catalog and in the sporting press, concentrated on the Models Four and Six. Initially the Model Four cost $449.95 and the Model Six, $399.95. Both models were short-lived as they were dropped from the 1988 catalog. It is interesting to note that internally the factory referred to both models as BDL grade.

The Models Four and Six featured positive-cut checkering on a slimmer pistol grip of the full cheek piece Monte Carlo buttstock, as well as on the flared forearm. This was a major change from the pressed basketweave checkering of the Model 742 and 760 BDL grade rifles. The step in the rear of the BDL receiver also was eliminated. A new cosmetic touch, reminiscent of earlier rifles, was a cartridge head medallion, denoting the caliber, which was inlaid in the barrel extension at the bottom of the receiver, forward of the magazine.

The model designation, *MODEL FOUR* or *MODEL SIX*, was stamped on the right side of the receiver. The left receiver

Front and back of brochure sent to dealers promoting the newly introduced Bullet Knife and the Model Four and Six Rifles. *Courtesy of Jack Heath .*

| Remington Model Four centerfire autoloader rifle introduced in 1981 *Courtesy of Remington Arms Co.* | Remington Model Six centerfire pump action rifle introduced in 1981. *Courtesy of Remington Arms Co.* |

Flier advertising pewter grip cap inserts for the Models Four and Six. *Courtesy of Jack Heath.*

panel was stamped *Remington* and the serial number was stamped between the fire control retaining pins. Both models shared a new block of serial numbers starting at A4000000. The barrel legend was stamped in two lines on the left side of the barrel above the forearm:

1981 – 1983:
REMINGTON ARMS CO. INC.
ILION, NY. MADE IN U.S.A.

1984 – 1988:
WARNING – READ INSTRUCTION BOOK FOR SAFE OPERATION – FREE FROM *****REMINGTON ARMS COMPANY, INC., ILION, N.Y., U.S.A.*****

The caliber was stamped after the barrel legend.

The Model Four autoloader rifle was offered in six calibers: .30-06, .308 Winchester, 7 mm Express Remington, .270 Winchester, .243 Winchester and 6 mm Remington.

The Model Six pump action rifle was offered in five calibers: .30-06, .308 Winchester, .270 Winchester, .243 Winchester and 6 mm Remington. It was not chambered for the 7 mm Remington Express cartridge.

In 1983 caliber 7mm Express Remington returned to its original name — .280 Remington. Thus far, a Model Four marked .280 REM has not been observed. Caliber .308 Winchester was dropped from the list of cartridges available in 1985 and

Remington Arms Company tie clasp, 2-7/8 inches long, in the shape of a Model Four rifle. *Author's collection.*

65

Walnut finish presentation box with all five Sid Bell-designed pewter grip cap inserts, as offered to dealers for display. The inside of the lid is marked "Reminton Model Four and Six Rifle Grip Cap Inserts. Solid Pewter by Sid Bell." *Courtesy of Jack Heath.*

engraving. Suggested retail price was $9.95.

The third tie-in was the "Collectors' Limited Edition" Model Four rifle, celebrating the 75th or diamond anniversary of the autoloading rifle. See the chapter on Commemorative Rifles for more details on the Diamond Anniversary Limited Edition Model Four.

Carbine versions of the Model Four & Six were discussed in late 1981 and prototypes were tested. However, they were never put into production.

Eight Model Fours in caliber .25-06 and another eight in caliber 7mm-08 successfully completed accuracy and function trials in late 1981, but the two versions were not added to the list of available calibers.

A total of five high grade engraved Model Fours and Model Sixes were made.

A total of 68,085 Model Fours, including the Limited Edition and two engraved models, were made. Production of the Model Six totaled 36,236, including three engraved models.

the 6mm Remington chambering was dropped in 1987. About 1,500 Model Fours and fewer than 1,000 Model Sixes were chambered for the 6 mm Remington.

The 1982 catalog introduced three unique promotional tie-ins for the Model Four and Six. The Bullet Knife R1123 returned to the Remington list of accessories and dealers could purchase one for each Model Four or Six sold. The dealer's price was $22.00 and the suggested retail price was $45.00. One blade was etched with a Model Four & Model Six inscription. Few were sold and this has become one of the most-desired modern Remington Bullet Knives. The original knife promotion poster is also very collectible.

A second promotion was the Sid Bell-designed pewter grip cap insert in four big game designs and one blank for

Table 1

1981 — 1988 Cataloged Models

Model	Caliber	Order Number	Production % of total
Four	.30-06	4748	49.2
Four Ltd. Ed	.30-06	4749	2.4
Four	.270 Win.	4744	23.9
Four	.243 Win.	4742	8.6
Four	.308 Win.	4750	8.4
Four	7mm Exp Rem.	4746	5.1
Four	6mm Rem.	4740	2.4
Six	.30-06	4686	50.2
Six	.270 Win.	4684	29.0
Six	.243 Win.	4682	10.3
Six	.308 Win.	4688	7.6
Six	6mm Rem.	4680	2.9

The Model Four, order number 4747, caliber 7mm Express Remington with a 2 shot magazine, was a special order for the European market. It is included in the above totals.

Table 2
Model Four and Six Cataloged Specifications

Barrel length	22 inches
Overall length	42 inches
Weight	7-1/2 pounds
Monte Carlo stock	
Length of pull	13-5/16 inches
Drop at comb	1-11/16 inches
Drop at Monte Carlo	1-13/16 inches
Drop at heel	2-1/2 inches

The 1987 catalog noted two different stock dimensions for both Model Four and Model Six: drop at comb three-eighths of an inch and drop at heel 1-3/16 inch. The author has no explanation for the variation in stock specifications. So far all Model Fours and Sixes examined have the same specifications as when introduced.

Chapter 8 —
Sportsman 74 and 76

The Sportsman series, conceived in mid-1983 as a way to increase Remington's market share, within three years, included 12 gauge autoloader and pump action shotguns, the 78 bolt action center fire rifles, 581-S bolt action rimfire rifles, the 74 autoloader center fire rifles and the 76 pump action center fire rifles. All of these firearms were lower-cost versions of models in the regular line and were promotional items directed toward the mass merchandisers.

The series, introduced in mid-1984 with its own brochure, did not have a suggested retail price. Generally, dealers and mass merchandisers sold the autoloader 74 in the $350 range and the pump action 76 in the $315 range. They were adver-tised in Remington's 1985 - 1987 regular catalogs and price lists as a separate line. The Sportsman 74 and 76 were officially discontinued December 31, 1987. The 1988 Remington catalog dropped the Sportsman series, including the 74 and 76, and only the Model 78 bolt action and Model 581-S bolt action rimfire rifles remained as part of the regular line. However, sales of the Sportsman 74 and 76 did continue into 1988 to clear out the warehouse.

The 74 and 76 rifles featured a plain, uncheckered, straight-comb, walnut-finished hardwood stock with a lacquer finish. All metalwork was matte finished, which eliminated three polishing steps, and the receiver was drilled and tapped for scope

1984 Sportsman Series introduction brochure. Author's collection.

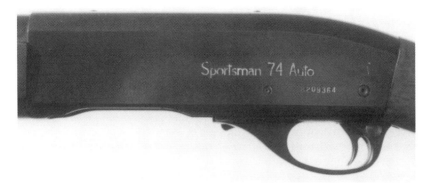

Sportsman 74 left side receiver markings. *Photo: Author. Courtesy of Richard VanDuesen Jr.*

mounts. The trigger guard was powder-coated and a less expensive bolt action Model 788 style rear sight was used.

The receiver was stamped in plain letters *Sportsman* and the model designation *74 Auto* or *76 Pump* was on the left side over the serial number. The barrel legend was stamped in two lines on the left side of the barrel above the forearm :

WARNING – READ INSTRUCTION BOOK FOR SAFE OPERATION – FREE FROM

***** REMINGTON ARMS COMPANY, INC., ILION, N.Y., U.S.A. *****

The caliber .30-06 SPRG or .280 REM was stamped after the barrel legend.

The Sportsman 74 and 76 were similar in style to the 1950s "A" grade Models 740 and 760.

Sportsman 74 and 76 rear sight. *Photo: Author. Courtesy of Richard VanDuesen Jr.*

Table 1
1984 – 1987 Sportsman Cataloged Models

Model	Caliber	Order Number	Production
74	.30-06	6166	47,881
74	.280 Remington	6170	1,574
76	.30-06	6168	20,715

The Sportsman 74, order number 6170, caliber .280 Remington with a 2 shot magazine, was a European market special order.

Table 2
Sportsman Cataloged Specifications

Barrel length	22 inches
Overall length	42 inches
Weight	7-1/2 pounds
Stock	
Length of pull	13-3/8inches
Drop at comb	1-11/16 inches
Drop at heel	2-1/8 inches

These are the same as the Model 7600.

Chapter 9 —
High Grade Rifles

The autoloader and pump action rifle receiver panels invite the artistry of the engraver. Remington master engraver Robert P. Runge best described these works of art and the relationship of the engraver to his work in Kevin McCormack's article, "When I Paint My Masterpiece," published in the autumn 2000 issue of *The Double Gun Journal*. "In general, the craft of engraving tends to demand a very focused, intense bond between the engraver and his work .... There is a part of me in every game scene and wedge of scrollwork for the world to enjoy, and the ability and the opportunity to leave that for the world has meant a great deal to me."

Remington factory engraved rifles do not have their left receiver panel roll marked with *Remington*, the model designation and serial number like the production rifles. Receivers for factory engraved rifles were assigned a serial number and removed from the production line before the roll markings were applied. The bottom of the breech ring/barrel extension was hand engraved *Remington,* and the receiver

Model 740 D grade, serial number 63681, left and right receiver panels with English style scroll engraving. *Photo: Author. Courtesy of Tom Cameron and Art Lewis.*

bottom was hand engraved with the model number and engraving grade over the serial number.

Some early factory engraved rifles had the model number, engraving grade and serial number hand engraved on the left receiver panel. One example observed by the author had *MODEL 740, PREMIER F GRD 79582* hand engraved on the left receiver panel and *Remington* was hand engraved on the bottom of the breech ring/barrel extension.

Stocks and forearms were furnished in almost any dimension required, and the high grade walnut was finely checkered by hand. All metal parts were carefully fitted and polished.

Remington's Custom Shop offered three grades of engraved firearms.

D - Peerless Grade

The receiver sides, top, bottom, barrel, barrel extension, trigger guard, magazine bottom and fore-end cap of the D - Peerless Grade were embellished with hand-cut English-style scroll engraving.

Model 740 D grade, bottom view showing hand engraved Remington over Model 740-D over serial number 63681, all outlined with English style scroll engraving. *Photo: Author. Courtesy of Tom Cameron and Art Lewis.*

REMINGTON MODEL 742 F PREMIER
CENTER FIRE AUTOLOADING RIFLE

Model 742 F grade, serial number 246183, left receiver panel with English style scroll engraving and three game scenes. Engraved 1967 by Robert P. Runge. *Photo: Remington Arms Co.*

Model 760, F grade with inlays, serial number A6958800, right and left receiver panels with English style scroll engraving and three game scenes inlaid in gold. Engraved 1976 by Robert P. Runge. *Photo: Author. Author's collection.*

Selected prices were:

Year	M740/M742	M760
1955	$509	$488.60
1965	$575	$575
1975	$880	$880
1978	$1,200	$1,200

F-Premier Grade

Each receiver panel in the F-Premier Grade had three game scenes and fine English-style scroll engraving. The same scroll engraving covered the receiver top, bottom, barrel, barrel extension, trigger guard, bottom of magazine and fore-end cap. Selected prices, by year, include:

Year	M740/M742	M760
1955	$925.50	$904.50
1965	$1,050	$1,050
1975	$1,550	$1,550
1978	$2,400	$2,400

F-Premier Grade with gold inlays

The top grade offered, the F-Premier Grade with gold inlays, had the same specifications as F-Premier grade rifles, but the three game scenes on each panel were inlaid in gold. The gold inlays increased the price by approximately 50 percent over F-Premier grade. Selected yearly prices were:

Year	M742	M760
1965	$1,750	$1,750
1975	$2,400	$2,400
1978	$3,600	$3,600

HIGH GRADE PRODUCTION 1952 – 1980

The factory engravers during this period were:

• Leo Bala, who performed "D" grade scroll work, starting in 1967.

• Jack H. Caswell, who started in 1969.

• Carl Ennis, a master

73

Model 760 F grade, serial number A6958800, close up of left receiver panel detailing Robert P. Runge's initials and year in the lower part of the deer scene. Initials and year are highlighted for clarity. *Photo: Author. Author's collection.*

engraver who had worked at Parker Brothers in the 1930s. He came to Ilion, NY when Remington purchased the company and moved the operation. Ennis retired in 1978.

• Robert P. Runge, a master engraver who also had worked at Parker Brothers in the 1930s. He came to Ilion when Remington moved the operation there. He retired in 1979.

The total number of "D" and "F" grade autoloader and pump action rifles completed during this period were:

Model 740 - 35
Model 742 - 75
Model 760 – 82

A complete description by serial number and grade of each high grade rifle is not available. However, some records, such as

Model 760 F grade, bottom view showing hand engraved Remington over Model 760-F over serial number A6958800. *Photo: Author. Author's collection*

74

Engraver's "pulls" of the right and left receiver panels of the Major General James B. Middleton Model 742 F grade rifle, serial number A710717. Engraved by Robert P. Runge. *Photo: Author. Courtesy of Larry "Babe" Del Grego.*

work books, engraver's "pulls", work order tags as well as observed rifles, provided the partial listing tabulated in the following tables. In some cases where the caliber was not noted in the original records, it was most likely .30-06.

Normally, Remington engravers did not sign their work. High grade rifles attributed to Carl Ennis were based on his work books from 1969 through 1978. In addition to high grade rifles, Ennis also engraved a number of arms with samples of scroll patterns and game scenes for review by the Remington Arms Company Operations Committee. These rifles were considered as possible candidates for roll marking production rifles.

Ennis also engraved a

Engraver's "pulls" of the left and right receiver panels, each with two bears, of a standard grade Model 742, serial number A6949955, done for Jim Martin, Manager, Product Design, Remington Arms Company. The trade name, model number, Remington and serial number were roll marked before Carl Ennis engraved the bears. *Photo: Remington Arms Co.*

75

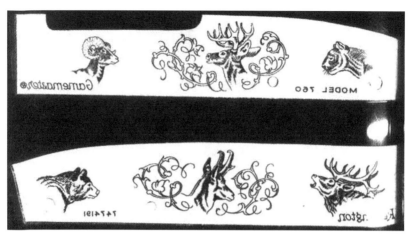

Engraver's "pulls" of the right and left receiver panels on a standard grade Model 760, serial number 7474191, done for review by the Remington Arms Company Operations Committee as a possible roll marked game scene. Engraved by Carl Ennis. Ennis engraved a number of "Product Improvement" game scenes over the years. *Photo: Remington Arms Co.*

number of "retirement" guns for retiring Remington employees. Many of the "sample" and "retirement" firearms were production items with the roll marked *Remington*, model number and serial number on the left receiver panel. They were not included in the following tables.

The listing of Robert P. Runge engraved "F" grade Model 760s was taken from a February 26, 1982 memorandum by Runge to Larry Goodstal, Curator of the

Remington Arms Company Museum. The memorandum covered the "F" grade Model 760s Runge engraved between 1953 and 1979. The "F" grade Model 740 and Model 742 rifles attributed to Runge are from his

Red work order tag for Model 760 F grade, serial number A6958800, order number AM 86375. Back of tag has Runge's penciled dates and hours of work, he spent 81.0 hours over 12 days from February 26, 1976 to March 17, 1976 to engrave and gold inlay this rifle. *Photo: Author. Courtesy of Larry "Babe" Del Grego.*

Robert P. Runge, Remington master engraver, at work. *Photo: Remington Arms Co.*

engraver's "pulls" and red work order tags in the Larry "Babe" Del Grego collection. The back of the work order tags have Runge's dates and hours worked for that rifle. On one engraver's "pull" he also noted the amount of gold wire used. Runge spent over 80

Table 1			
Model 740 High Grade Rifles			
Serial #	*Grade*	*Caliber*	*Notes*
1001	F gold inlays	.30-06	First production Model 740, Remington President C. K. Davis rifle, C. Ennis engraved
46539	F gold inlays	.30-06	Left hand safety, R. P. Runge engraved
46540	D	.30-06	
46541	D	.30-06	
57311	F gold inlays	.30-06	R. P. Runge engraved
57312	F gold inlays	.30-06	R. P. Runge engraved
63679	D	.30-06	Low comb stock
63681	D	.30-06	Special stock
63682	D	.308 Win.	
63684	D	.30-06	
79582	F gold inlays	.30-06	

Don Talbot, current Remington master engraver, at work. *Photo: Remington Arms Co.*

hours to engrave and gold inlay a rifle. He used 74-1/2 inches of .030 gold wire and 14-3/4 inches of .040 gold wire for the inlays on Model 742 F grade serial number 223798.

Only two high grade Model Four and three high grade Model Six rifles were completed.

An F grade Model 7400 autoloader with special serial number 1816 is currently displayed at the Remington Arms Company, Ilion, NY factory.

Current production Model 7400 autoloaders and Model 7600 pump action rifles are still available with D & F grade engraving. Checkering is hand cut at 22 lines per inch in "D" grade and 24 lines per inch in "F" grade. Five different grip caps

and four different buttplates or recoil pads are available.

A 1994 Custom Shop brochure detailed the services available and the prices:
- Grade D — $2,509
- Grade F — $5,169
- Grade F with gold inlay — $7,752

The 1997 Custom Shop brochure increased the prices to:
- Grade D — $2,610
- Grade F— $5,377
- Grade F with gold inlay — $8,062

The 2001 prices were:
- Grade D — $2,890
- Grade F — $5,953
- Grade F with gold inlay — $8,924

Don Talbot is the current Remington factory engraver.

Table 2
Model 742 High Grade Rifles

Serial #	Grade	Caliber	Notes
28943	D	.308 Win.	
28944	D	.30-06	Scope, four magazines
28945	D		
28946	F	.30-06	
159687	D	.308 Win.	
181467	D	.308 Win.	
207661	D	.308 Win.	
207662	D	6mm Rem.	Recoil pad
211239	D	.30-06	
211240	D	.30-06	
211241	D	.30-06	
211242	D	.30-06	
211243	D	.280 Rem.	Monte Carlo stock & sling
223798	F gold inlays	.30-06	R. P. Runge engraved
226341	F	.280 Rem.	
226342	D	.30-06	Duplicate serial number, shipped 1966
226342	D	.280 Rem.	Duplicate serial number, shipped 1967
246182	D	.280 Rem.	
246183	F	.308 Win.	R. P. Runge engraved
246185	F	.30-06	R. P. Runge engraved
302921	F	.30-06	
302922	D	.308 Win.	Monte Carlo stock.
313106	F gold inlays	.308 Win.	R. P. Runge engraved
332062	D	.30-06	
332063	F	.30-06	Scope. R. P. Runge engraved
351862	D	.280 Rem.	
351863	F	.30-06	R. P. Runge engraved
384826	F gold inlays	.30-06	R. P. Runge engraved game scenes and gold inlays; scrollwork by J. Caswell
390168	F gold inlays	.30-06	R. P. Runge engraved game scenes and gold inlays; scrollwork by L. Bala
390169	F gold inlays	6mm Rem.	

A1000000	F	.30-06	Special serial # for one millionth Model742, made December 1976. Displayed in Remington Museum. R. P. Runge engraved
A6902981	D	.30-06	
A6978123	D	.30-06	
A7017882	F	.30-06	R. P. Runge engraved
A7038474	D	.30-06	
A7043011	D	.30-06	
A7092827	D	.30-06	Carbine
A7104072	D	.30-06	
A7118812	D	.30-06	Left handed stock
A7143573	F gold inlays	.30-06	Russian Embassy order. R. P. Runge engraved game scenes, gold inlays; scrollwork by J. Caswell.
A7143574	F gold inlays	.30-06	Russian Embassy order. R. P. Runge engraved game scenes, gold inlays; scrollwork by J. Caswell.
A7143575	F gold inlays	.30-06	Russian Embassy order. R. P. Runge engraved game scenes, gold inlays; scrollwork by J. Caswell.
A7143576	F gold inlays	.30-06	Russian Embassy order. R. P. Runge engraved game scenes, gold inlays; scrollwork by J. Caswell.
A7194444	D	.30-06	
A7202233	D	.30-06	
A7212778	F gold inlays	.30-06	R. P. Runge engraved; "special scenes"
A7266031	F gold inlays	6mm Rem.	R. P. Runge engraved
A7272489	D	.308 Win.	
A7338840	F		C. Ennis engraved
A7343970	F		C. Ennis engraved
A7343971	F		C. Ennis engraved
A7343972	F		C. Ennis engraved
A7393672	F gold inlays		R. P. Runge engraved
A7496180	F gold inlays	.30-06	
A710715	F gold inlays		R. P. Runge engraved
A710717	F SPEC		Maj. Gen. James Middleton order, bridge & cannons on left receiver, R. P. Runge engraved

Table 3
Model 760 High Grade Rifles

Serial #	Grade	Caliber	Notes
1001	F gold inlays	.30-06	First production Model 760, Remington President C. K. Davis rifle, C. Ennis engraved
2736	F		C. Ennis engraved
123700	F		R. P. Runge engraved
123701	D	.30-06	
208847	F		R. P. Runge engraved
208849	F		R. P. Runge engraved
208852	F		R. P. Runge engraved
208853	F		Dewey Godfrey, Remington Vice President & Director of Sales rifle; R. P. Runge engraved
244794	F	.30-06	
244795	F		R. P. Runge engraved
244796	F		R. P. Runge engraved
255365	F gold inlays	.270 Win.	R. P. Runge engraved
255367	F	.30-06	R. P. Runge engraved
255368	F		R. P. Runge engraved
255369	F		R. P. Runge engraved
255490	F		R. P. Runge engraved
255492	F		R. P. Runge engraved
255494	F gold inlays		R. P. Runge engraved
352063	D	.308 Win.	
352064	D	.308 Win.	
424329	D		C. Ennis engraved
430670	D	.35 Rem.	Carbine
446041	D	.308 Win.	
459533	D	.308 Win.	
494403	F	.30-06	R. P. Runge engraved
519291	D	.243 Win.	
A6958800	F gold inlays	.30-06	R. P. Runge engraved
A6958801	F gold inlays		R. P. Runge engraved
A6978124	D	.30-06	
A7065702	D	.30-06	
A7107912	F gold inlays	.270 Win.	R. P. Runge engraved
A7206786	D	.30-06	
A7206787	D	.30-06	
A7295254	F gold inlays	.30-06	R. P. Runge engraved
A7349194	F		C. Ennis engraved
A7468635	F gold inlays		R. P. Runge engraved
A7496182	F gold inlays		R. P. Runge engraved
A710719	F gold inlays		Last R. P. Runge engraved rifle, November 27, 1979

Chapter 10 —
Commemorative Rifles

Remington commemorative firearms were issued in 1966 to celebrate the company's 150th Anniversary. The 150th Anniversary Models, in caliber .30-06, had the left receiver panel roll marked with a gold-filled Remington 150th Anniversary logo. A total of 11,412 Model 742s and 4,610 Model 760s 150th Anniversary Models were made. Prices were $149.95 for the Model 742 and $134.95 for the Model 760.

The Canadian Centennial Model 742 in caliber .308 Winchester was manufactured in 1966 and sold in 1967. The left receiver panel was roll marked with a Canadian Centennial logo that was gold-filled, and the right side of the buttstock had an inlaid Canadian Centennial 1867 – 1967 medallion.

Cased sets consisting of a Remington Canadian Centennial Model 742 and a similarly appointed Ruger 10/22 autoloader in caliber .22 rim fire were sold in 1967. Some 1,968

Canadian Centennial Model 742s were sold to a distributor in Canada, Peterborough Guns. There was no retail price published in the United States.

Remington celebrated the 1776 - 1976 United States Bicentennial with rifles and shotguns that carried the Bicentennial logo. Both the Model 742 and the Model 760, in caliber .30-06, had the left receiver panel roll marked with a gold-filled Bicentennial logo and the years 1776 — 1976. Some 10,108 Bicentennial Model 742s at $244.95 each and 3,804 Bicentennial Model 760s at $214.95 were sold.

In 1981 Remington introduced its first Limited Edition firearm — a Model 1100 autoloader shotgun commemorating over 75 years of building autoloader shotguns. The second Limited Edition was a Model Four autoloader rifle that commemorated 75 years of autoloader rifle production. It was issued in 1982 at a price of $1,500.

Model 760 left receiver panel with roll marked 150th Anniversary logo highlighted in gold color. *Photo: Author. Author's Collection.*

Model 742 left receiver panel (above) with roll marked Canadian Centennial logo highlighted in gold color, and right side of buttstock (right) with inlaid Canadian Centennial medallion. *Photo: Roy Marcot. Courtesy of Remington Arms Co.*

Both side panels of the Model Four receiver were etched with scrollwork and the left panel featured a hunting scene based on a N. C. Wyeth painting from the Remington Arms Company Collection. The right panel featured the Model 8 and Model Four autoloader rifles. The hunting scene, rifles and Limited Edition logo were gold plated. The high grade walnut straight comb buttstock and flared forearm featured cut checkering and a satin finish. The barrel legend read:

CUSTOM BUILT BY
REMINGTON ARMS CO.
—— ILION, NEW YORK ——

The caliber ".30-06" was stamped where barrel date code would normally have been stamped. A .30-06 cartridge head was inlaid in the bottom of the barrel extension.

The serial numbers were in a special series starting with LE81-0001. Approximately 1,400 rifles were shipped from 1982 to 1988. Most were packaged in a white box with a white end label reading: MODEL FOUR LIMITED EDITION – ONE OF FIFTEEN HUNDRED. The packer and date code, as well as the order number, 4749, were stamped on the end label. The serial number was handwritten. A cream colored sleeve with Limited Edition artwork and a description of the model completed the packaging.

Later shipments used

Model 742 left receiver panel with roll marked Bicentennial logo done in gold color. *Photo: Author. Courtesy of Remington Arms Co.*

Remington's standard green and gold end label that was stamped M/4 LIMITED EDITION, along with the packer and date codes and the 4749 order number.

Besides the standard packing inserts, a special Model Four Limited Edition Certificate Application was included. When the application was returned to Remington,, a certificate of ownership suitable for framing was sent and the owner's name was entered in a registry.

A separate series of around 100 Limited Edition rifles were shipped in 1994 to finish up existing etched and plated receivers. These rifles were based on the Model 7400, as the Model Four had been discontinued in 1988. The barrel extension lacked the inlaid cartridge head and the stocks were standard grade, as were the grip cap and buttplate. The barrel had the standard Model 7400 proofs, assembler stamps, date code and barrel legend. The "B" prefix serial number was part of the standard Model 7400 and Model 7600 serial number series. The rifles were shipped in Remington's standard green box

End label of Limited Edition shipping box (left) with description of rifle, hand written serial number LE81-1234, order number 4749 and packer's code. The packer's code, in the lower left-hand corner, has the packer's numerical code and the date code. The date code WD indicates assembly August 1983. (Below) Model Four Limited Edition's special shipping sleeve. Photos: Author. Author's Collection.

with the end label stamped LIMITED EDITION, order number 4749, "B" prefix serial number and the packer and date codes. At least one late production Limited Edition rifle was shipped without gold plating. It was chambered for the .308 Winchester cartridge.

One Limited Edition rifle, destined for the Remington Museum Collection, was issued a special serial number — LE81-1816.

The 1991 catalog featured Remington's 175th Anniversary scroll design with an American Eagle roll marked on the left receiver panel of Model 11-87 shotguns and Model 7400 rifles. Approximately 5,000 cataloged caliber .30-06 Model 7400 175th Anniversary rifles were sold. Two non-cataloged 175th

Anniversary models, special ordered by distributors, included a Model 7400 in caliber .270 Winchester and a Model 7600 in 7mm-08. Approximately 1,200 caliber .270 Winchester Model 7400s and 1,000 caliber 7mm-08 Model 7600s were sold. The non-cataloged Model 7600 was handled by Grice Wholesale of Clearfield, Pennsylvania. The rifle's standard Monte Carlo style buttstock and forearm featured cut checkering and a high gloss finish. The Model 7400 cost $515 and the Model 7600, $498.

In 1992 Grice Wholesale commissioned the non-cataloged B/O Special "Deer Hunter" Model 7600. The left receiver panel featured a gold-filled roll marked deer scene based on a Robert Kuhn painting in the

Remington brochure promoting the special features of the Model Four Limited Edition rifle. *Courtesy of Remington Arms Co.*

85

1991 catalog page - Model 7400 left receiver panel with roll marked 175ᵗʰ Anniversary logo. *Courtesy of Remington Arms Co.*

Remington Arms Company Collection. The standard Monte Carlo style buttstock and forearm featured cut checkering and a high gloss finish. Some 1,000 rifles were sold in caliber 7mm-08 Remington, 500 in caliber .270 Winchester, and 500 in caliber .30-06.

In addition, Limited Edition knives with a matching deer scene also were offered by Grice. Grice Wholesale continues to offer non-cataloged variations of the Model 7600; however, this was the only variation with a special roll marked scene.

The 1997 Remington catalog featured the "Buckmasters American Deer Foundation" Limited Edition Models 7400 and 7600, in caliber .30-06, at a price of $600 and $567 respectively. Both receiver panels had extensive pressure applied fine-line engraving and BUCKMASTER ADF LIMITED EDITION stamped in a banner. Less than 800 rifles of each model were sold. The catalog noted that in conjunction with Buckmasters, a donation to charity would be made for each ADF Limited Edition rifle sold.

The Remington 180ᵗʰ Anniversary Commemorative five gun set consisted of the Model 7400 autoloader rifle, Model 7600 pump action rifle, Model 700 bolt action rifle, Model

Grice Wholesale "Deer Hunter" Model 7600 left receiver panel with gold filled roll marked deer scene. Caliber 7mm-08 Remington. Rifle barrel date code is AM indicating assembly March 1992. *Photo: Author. Author's collection.*

Model 7400 left receiver panel (above) with BUCKMASTERS AMERICAN DEER FOUNDATION logo and pressure-applied fine-line engraving, and (below) right receiver panel with BUCKMASTERS ADF LIMITED EDITION banner and the same engraving.*Photos: Author. Courtesy of Remington Arms Co.*

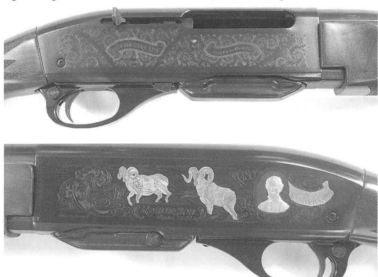

180th Anniversary Model 7400 left receiver panel (above) with gold embellished rams, bust of Eliphalet Remington and 1816 – 1996 ANNIVERSARY banner, and (below) right receiver panel with two gold embellished elk. *Photos: Blair Grabski. Courtesy of Blair Grabski.*

870 pump action shotgun and Model 1100 autoloader shotgun. The engraving and gold embellished game figures, 180th Anniversary 1816 – 1996 banner and bust of Eliphalet Remington were done in Italy. Fewer than 30 sets were delivered in 1997. The Model 7400 180th Anniversary rifle had a pair of rams, 180th Anniversary banner and bust of Eliphalet Remington on the left receiver panel and a pair of elk on the right receiver panel. The Model 7600 180th Anniversary rifle had a bear, 180th Anniversary banner and bust of Eliphalet Remington on the left receiver panel and a pair of deer on the right receiver panel. The name of the engraver – Bottega C. Giovanelli – appears on the left receiver panel of both rifles. Serial numbers started at 7400ER001 for the Model 7400 and 7600ER001 for the Model 7600.

A privately-commissioned B/O Special "Larry Benoit" Model 7600 Carbine, caliber .30-06, was offered by Wilderness Trading & Supply Co. of Marshfield, Vermont in 1999 and 2000. The Larry Benoit Commemorative honored a native Vermonter nationally known as one of the greatest whitetail deer hunters in the country. *Sports Afield* had cited him as possibly the best deer hunter in America. Larry's favorite rifle is the Remington pump action carbine thus the carbine based commemorative.

The left panel had the standard Remington pressure-

180th Anniversary Model 7600 left receiver panel (above) with gold embellished bear, bust of Eliphalet Remington and 180th 1816 – 1996 ANNIVERSARY banner, and (below) Model 7600 right receiver panel with two gold embellished deer. *Photos: Blair Grabski. Courtesy of Blair Grabski.*

Model 7600 Carbine with the Larry Benoit's biggest buck, signature, 1999 and "One of One Thousand" etched and gold plated. *Photo: Author. Courtesy of S. Scribner, Wilderness Trading and Supply Co.*

applied fine-line game scene. The right side of the receiver had a head mount of Larry's biggest buck, year and Larry's signature over "One of One Thousand" etched and plated in gold. Series 1999 carbines were numbered LEB 0001 to LEB 0110.

The second delivery of the Series 1999, serial numbers LEB 0061 to LEB 0110, as well as all Series 2000 carbines, had the new Remington Integrated Security System (ISS) trigger lock that was part of the safety.

Series 2000 carbines were numbered LEB 0111 to LEB 0170.

A total of 170 Larry Benoit Commemorative Model 7600 carbines were made at a price of $995. The end label carried the standard carbine order number 4661, special serial number beginning with LEB plus a small red label noting B/O SPECIAL.

Gold-filled roll marked receivers with worn gold fill can be repaired with the Bonanza Gold Fill Kit from Forester Products, Inc.

End label of Benoit Commemorative shipping box with description of rifle, serial number LEB 0132, order number 4661 SPL, date code and B/O SPECIAL label. The packer's code is partially covered by the RGC 4661 SPL sticker. The 29PU date code indicates assembly 29 June 2000. *Photo: Author. Author's collection.*

Commemorative Rifle Production

Year	Model	Order No.	Caliber	Description	Number Produced
1966	742		.30-06	Remington 150th Anniversary	11,412
1966	760		.30-06	Remington 150th Anniversary	4,610
1967	742		.308 Win	Canadian Centennial	1,968
1976	742	7612	.30-06	United States Bicentennial	10,108
1976	760	7614	.30-06	United States Bicentennial	3,804
1982	Four	4749	.30-06	Autoloader 75th Anniversary App.	1,500
1991	7400	4700	.30-06	Remington 175th Anniversary App.	5,000
1991	7400	4701	.270 Win.	Remington 175th Anniversary App.	1,200
1991	7600	4672	7mm-08	Remington 175th Anniversary App.	1,000
1992	7600	4666*	.270 Win	Grice "Deer Hunter Special"	500
1992	7600	4668	7mm-08	Grice "Deer Hunter Special"	1,000
1992	7600	4670*	.30-06	Grice "Deer Hunter Special"	500
1997	7400	4695	.30-06	Buckmasters ADF	fewer than 800
1997	7600	4669	.30-06	Buckmasters ADF	fewer than 800
1997	7400	5132	.30-06	Remington 180th Anniversary	fewer than 30
1997	7600	4675	.30-06	Remington 180th Anniversary	fewer than 30
1999	7600C	4661*	.30-06	Larry Benoit Special	100
2000	7600C	4661*	.30-06	Larry Benoit Special	70

* Same as regular order numbers.

Chapter 11 —
Unusual Rifles

Several "Featherweight Model 760 rifles were built in late 1952 and early 1953 in response to rumors that the competition was about to introduce a 6-pound pump action rifle. This was accomplished by using an aluminum alloy receiver to reduce the weight of the Model 760 rifle. Six "Lightweight"(as they were now called) aluminum alloy receiver rifles were field tested and sales representatives, in a September 3, 1953 Operations Committee – Arms Division report, requested a total of 15 rifles for demonstration purposes. After review, the Committee sent Remington's management an inquiry seeking approval for further prototype production. A September 15, 1953 reply advised "that the use of aluminum receivers for high-powered rifles is considered unacceptable by Management." All further work with aluminum alloy receivers for high-powered rifles was dropped.

Two Model 760s with aluminum alloy receivers have been observed. Serial number 104881, in the Remington Arms Co. Collection, has only the receiver, forearm and buttstock. The action tube assembly has been drilled to reduce weight.

Serial number 104877, from the Clyde Dora collection, is an ADL grade rifle in caliber .270 Winchester. Its barrel date code is June 1953 and its action tube assembly is not drilled for lightness. The anodizing on the aluminum alloy receiver is slightly darker than the blued barrel, and the end plug in the action tube assembly also is made from anodized aluminum alloy.

Model 760, ADL grade, serial number 104877, has an experimental aluminum receiver. The caliber .270 Winchester barrel carries a barrel date code of June 1953. The receiver is drilled and tapped for scope and receiver sights. *Photo: Author. Courtesy Clyde Dora.*

Factory Cutaway Rifles

Remington, over the years, has made a number of factory cutaways to illustrate the operation of the autoloader and pump action rifles. The 1989 Jack Appel auction had a cutaway Model 742, serial number 10023, that was used to illustrate the "power-matic" gas system in period catalogs and brochures.

John Lacy, author of *"The Remington 700"*, quotes a factory letter noting that in 1968 each of the eight Regional Offices received the following eight cutaways:

• M700 bolt action Varmint Special, caliber .22-250
• M742 autoloader, caliber .30-06
• M742 autoloader Carbine, caliber .30-06
• M788 bolt action, caliber .222 Remington
• M788 receiver and bolt
• M581 bolt action, caliber .22 rimfire
• M581 receiver and bolt
• M1100 autoloader shotgun receiver.

Model 742 rifle, serial number 396549, from the author's collection, is one of these cutaways. This rifle was assembled in November 1968 as the last Model 742 serial number in the original series was 396562 assembled Nov. 26, 1968. After this date a new serial number range starting with A6900000 was used for both Models 742 and 760. The rifle is cut away over the chamber area, right side of the receiver and right side of the forearm over the gas operation system.

The factory collection includes a Model 742, caliber .30-06, serial number B6939909, with four cutaways in the right receiver panel.

Factory Model Stamping Error

The author's collection includes a Model 742, in caliber .30-06, serial number A7461226, barrel date code October 1977, that is stamped on the left receiver panel: *Gamemaster*, MODEL 760. It managed to elude at least four or five inspections, including the final packing verifying the serial number.

Wildwood Inc. of China, Maine has a Model 742, in caliber .308 Winchester, serial number 7394878, barrel date

Factory cutaway Model 742, serial number 396549, one of eight cutaway Model 742s done for Remington Regional offices. *Photo: Author. Author's collection.*

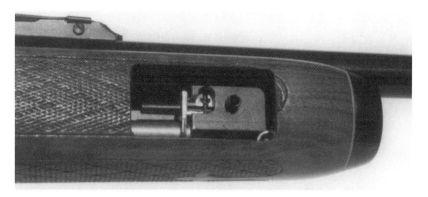

Factory cutaway Model 742 showing the opening in the fore-end exposing the gas operation system. *Photo: Author. Author's collection.*

code October 1973, that also is stamped on the left side: *Gamemaster, MODEL 760.*

Factory Assembly Error

The George Coldren collection includes a mint in box factory assembly error. It is a Model 742 receiver, serial number B7191096, marked *Woodmaster, MODEL 742,* assembled with a Model 760 barrel and pump action parts. The caliber is 6mm Remington. The end label of the box is stamped Model 760, Gamemaster, pump action rifle, Caliber 6mm Remington, serial number B7191096, order number 5900 and packer code of 3211 KV. The KV date code indicates an assembly date of May 1979. This rifle eluded four or five inspections including the final packing verifying the serial number.

Factory Caliber Marking Errors

6 mm Rem. Mag

John Lacy, author of *"The Remington 700,"* quotes a July 16, 1974 factory letter concerning a Remington 700 stamped 6MM REM. MAG. with the MAG. exed out. Apparently during the development of the new 6mm cartridge in 1962, it was first called the 6mm Remington Magnum before settling on 6mm Remington. A roll die for 6MM REM. MAG. was ordered and used on a few Model 700 bolt action and Model 742 autoloader rifles before the error was discovered and corrected. The MAG. was exed over before any rifles were shipped. The caliber 6MM REM. XXX has appeared on at least two Model 742s, serial number 102124, barrel date code February 1963 and serial number 103493, barrel date code March 1963.

7MM Exp Rem / 280 Rem

In 1957 Remington introduced the .280 Remington in the Model 740 autoloader. It was continued in the Model 742. The name was changed to 7mm Express Remington in 1979.

By the 1983 catalog, it was being referred to as the .280 REM (7MM EXP REM) and the 1988 catalog called it just the

Model 742, serial number A7461226, is stamped on the left receiver panel Gamemaster, MODEL 760. The .30-06 barrel has a date code of October 1977. *Photo: Author. Author's collection.*

.280 REM. Some Model 700 bolt action rifles have been reported with a dual caliber marking - 7MM EXP. REM. 280 REM. It is possible a few Model 7400 or Model 7600 rifles may have the dual caliber marking, however, the author is not aware of any so stamped.

Chapter 12 —
Experimental Rifles

The Remington engineers have developed a number of interesting prototypes in their quest to improve and upgrade the center fire autoloader and pump action rifles. These are only a few of their experiments that were not put into production.

The Model 740 Military Modification was developed from a February 22, 1951 request to develop two shooting models for consideration by the Small Arms Division of the US Army Ordnance Corps. The Model 740 Military Modification was approved, two models were constructed and then informally tested. The Military Models had the operating handle moved to the action bar and used the standard Model 740 buttstock, as well as a forearm without finger grooves. No formal tests were held on these rifles.

During Remington's long history, the lever action rifle has not been a production item. However, in the early 1960s various trial models of such rifles were constructed. Sometime around 1961, a one-off conversion of a Model 760 pump action rifle to a lever action was completed and tested, but not put into production. In 1962 Remington advertised its first production lever action repeater — the Nylon 76 "Trailrider" in .22LR, but discontinued it in 1965.

A series of rifles, code-named XC, were started in September 1958; however, the project was terminated January 1963. The XC-6 autoloader, XC-7 lever action and XC-8 pump action rifles all used receivers that were approximately the size and shape of the 12 gauge Sportsman 58 shotgun receiver. The XC series were a modification of the German Gewehr 43 World War II autoloader rifle. The locking lugs were loosely set into each side of the bolt and a slide moved the lugs into contact with recesses in the receiver wall.

The autoloader XC-6 used a variation of the Gewehr 43 stationary gas piston and moving

Model 740 Military Modification number 1, caliber .30-06, no serial number. Operating handle has been moved to the action bar and the magazine release has been extended. Circa 1951. *Photo: R. Marcot. Courtesy of Remington Arms Co.*

XC-7 lever action receiver with Gewehr 43-style locking lugs. The receiver is approximately the size of the 12 gauge Sportsman 58 shotgun receiver. *Photo: R. Marcot. Courtesy of Remington Arms Co.*

gas cylinder, which placed the push rod slightly below the center of the barrel. The lever and pump action versions manually moved the bolt and slide. Functioning models were developed, but the project was dropped as the rifles offered no significant advantage over the Model 742 and 760.

The introduction of the Browning BAR autoloading rifle, capable of handling magnum cartridges, resulted in attempts to upgrade the Models 742 and 760 to handle the 7mm Remington Magnum and .300 Winchester Magnum cartridges. The Models 744 and 766, as the improved versions of the Model 742 and 760 were called, began development in early 1969, but were discontinued in late 1970 in favor of another project --—the Model 732.

The Model 732 project began in early 1970 and was terminated in late 1974. The model designation was changed to Model 742X around late 1971 or early 1972. The final result was a detachable box magazine rifle capable of handling magnum cartridges. The rifle broke open like the Belgian FN-FAL and the US M16. The rotary multiple lug bolt traveled into a recess in the buttstock to minimize the receiver length.

Advantages of the break-open design were ease of cleaning, assembly, bore sighting and the ability to interchange rifle barrels. Models were

Experimental Model 760 lever action rifle, serial number 363980, caliber .30-06. Circa 1961. *Photo: Courtesy of Remington Arms Co.*

96

developed and test fired; however, the program was terminated due to weight and cost considerations. A companion autoloader shotgun, designed along the same principles, was developed at the same time.

Section Manager Jim Martin was in charge of the break-open program. Design team members were Leonard Hagen, Jim Hutton and Ken Rowlands.

By 1974, Remington had begun a product improvement program for its Models 742 and 760 that cumulated in the "New Generation" rifles. The Models Four, Six, 7400 and 7600 were introduced in 1981.

As part of the program, Remington experimented with different embellishments of the receiver panels. This included high relief etching, as well as etched scrollwork combined with gold-plated game scenes. The latter was later used on the Model Four Limited Edition rifle.

Experimental Model 742X, break open action, original version. Circa 1970. *Photo: Courtesy of Remington Arms Co.*

Experimental Model 742X, break open action, revised version. Circa 1972. Leon Johnson did the stock work and the F grade engraving was by Carl Ennis. *Photo: Courtesy of Remington Arms Co.*

Model 742 receiver etched with an experimental high relief game scene. *Photo: Author. Courtesy of Remington Arms Co.*

Above and below: Model 7400 right and left receiver panels etched with scrollwork and gold-plated game scenes. *Photo: Author. Courtesy of Remington Arms Co.*

Chapter 13 — **Magazines**

The detachable box magazines, used in Remington's autoloader and pump action rifles, are a direct descendant of James Paris Lee's 1879 patent. Lee developed his original detachable box magazine at E. Remington and Sons during 1878, and patented it on Oct. 4th 1879. By 1884, J. P. Lee and two Remington employees, Roswell F. Cook and Louis P. Diss, had patented a number of magazine improvements. The success of their combined efforts can be measured by the fact that today's detachable box magazines use the same design elements.

Remington's autoloader and pump action detachable box magazine has a capacity of four cartridges. The autoloader version's follower holds the bolt open after the last shot. L. Ray Crittendon, in 1951, developed the magazine bolt release lever, which depressed the back of the follower to release the bolt. These features were not needed on the pump action model.

All Remington magazines have the same external dimensions.

Those magazines intended for use with cartridges in the .30-06 class have full length followers. Shorter cartridges have spacer or filler pieces in the front and rear of the magazine to make up the difference in length.

The much smaller diameter .222 Remington and .223 Remington cartridges have magazines that are narrowed in the middle, as well as having front and rear spacers.

The caliber designation was stamped on the right hand side of each magazine. Remington, in 1951, discussed stamping multiple caliber designations on the side of the magazine, but it was decided to only have a single cartridge designation stamped on the side to avoid error in loading the rifle. This policy held until the 1990s when multiple cartridge designations were stamped on the side. Magazines for both lengths of cartridges lacking any caliber designation have been observed.

The magazine had been revised, altered and improved a number of times since 1952, yet

Initial magazine introduced in 1952 with the Model 760. Correct for first year production Model 760s. *Photo: Author.* Author's Collection.

it still retained its basic shape and external dimensions. The introduction dates for magazine changes are approximate, as Remington uses up stocks of older magazines first.

Description of the Magazines

The initial magazine, introduced with the Model 760 in 1952, had flat sides with holes. The caliber was stamped on the right side. Two Model 760s, serial number 8068, in caliber .30-06 and serial number 21291 in caliber .300 Savage, had this style magazine. The purpose of the holes is unknown, but they may have been used to provide a visual count of the cartridges.

The follower, in the caliber .30-06 magazine, had a noticeable raised nib to retain the cartridge at the shoulder.

The .300 Savage and .35 Remington magazines lacked the raised nib and had filler pieces both front and rear to make up the difference in the cartridge length. Many of the early magazines were hand fitted to rifles, as file marks on the feed lips are visible.

The second version, introduced in late 1953 or early 1954 on the Model 760, retained the flat sides but eliminated the holes and changed feed lip dimensions. The magazine also was used in 1955 as the initial magazine with the autoloader Model 740 due to feeding problems with a ribbed magazine.

A few of the early magazines have a "2" engraved on the side. The meaning of the "2" is not

Second version magazine introduced in late 1953 or early 1954 with the Model 760. The side hole is dropped. Also correct for first year (1955) production Model 740s. *Photo: Author. Courtesy of R. Shields.*

Third version magazine introduced in late 1954 with the Model 760. The vertical side ribs control the forward movement of the cartridge. It wasn't until 1956 that vertical cartridge control ribs appeared on Model 740 magazines. *Photo: Author. Author's Collection.*

Fourth version magazine introduced in 1960 with the Models 742 and 760. The caliber designation is boxed. *Photo: Author. Author's Collection.*

known, but speculation suggests that they may be second magazines, either given out with early rifles or for rifles ordered with two magazines.

The third magazine version, introduced in mid to late 1954 on the Model 760, added a vertical rib on each side of the magazine to help retain the shoulder of the cartridge and prevent battering of the bullet tip against the front of the magazine. Early Model 740 magazines left the ribs off because of a cartridge feeding problem due to a "wedging" action on the cartridge shoulder. The raised nib on the follower was retained. The wedging problem was corrected and ribs appeared on the Model 740 magazines within a year. Remington purchased the rights to patent number 2,434,269, which covered the use of cartridge shoulder retaining ribs, from Earle Harvey in 1957. The June 15, 1957

addition of the .222 Remington cartridge to list of calibers available for the Model 760 required that the sides of the Model 760 magazine be narrowed to retain the smaller diameter cartridges. The .222 Remington magazine had the vertical ribs but dropped the raised nib on the follower.

The fourth version, introduced in 1960, was identified by a boxed caliber designation. It was found on the Model 742 and Model 760, and had a number of dimensions revised for better cartridge control. The two vertical ribs were continued in order to hold the cartridges from going forward. The follower nib was at first dropped and then reinstated at a reduced height. Cartridge length, as before, was adjusted by varying filler pieces both front and rear.

Fourth version magazine, European specification 2 shot. *Photo: Author. Author's Collection*

A two shot magazine, in 7mm Express Remington, was developed for the European market. The magazine was indented on the sides so that the follower would only move part way down the magazine well.

The Model 760 added the .223 Remington cartridge May 1964. The .223 Remington magazine used the same narrowed magazine and follower as the .222 Remington magazine; however, the rear filler piece was shorter to allow for the difference in cartridge length. The .223 Remington magazine had cartridge feeding problems that were not resolved and the caliber was dropped from the 1969 catalog listing for the Model 760.

The fifth version, with the caliber designation enclosed in a rounded end box, again revised feed lip geometry and dimensions. It was introduced around 1980, just before the Models 7400 and 7600.

The sixth magazine version appeared around 1987, and the caliber designation was in a box with pointed ends. The rear feed lips, especially on the full length cartridge magazine, were extended for better cartridge control. Remington later updated warehouse stocks of these magazines by electrically etching additional calibers below the original stamped single caliber marking.

The seventh version,

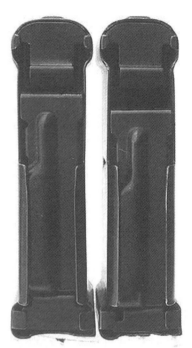

Left: caliber .222 Remington magazine. Right : .223 Remington Magazine. Both magazines have narrowed sides with front and rear filler pieces. Note that the .223 Remington follower is the same length as the .222 Remington.. *Photo: Author. Author's Collection.*

Fifth version magazine with caliber designation in a rounded end box. Correct for early Model Four, Six, 7400 and 7600 rifles. *Photo: Author. Author's Collection.*

Sixth version magazine with caliber designation in a pointed end box. The rear feedlips were extended for better cartridge control. Introduced around 1987. *Photo: Author. Author's Collection.*

Sixth version magazine with additional calibers electrically etched below the original stamping. *Photo: Author. Author's Collection.*

Seventh version magazine with multiple calibers electrically etched on the side. Introduced in the 1990s and updated as additional calibers were introduced. *Photo: Author. Author's Collection.*

introduced in the 1990s, had electrically etched multiple caliber listings on the side. This listing was revised as the list of calibers chambered in the autoloader and pump action rifles changed. The last magazines for the full length cartridges were etched .25-06, .270 WIN, .280 REM, .30-06 and .35 WHELEN.

The magazines for the shorter cartridges were etched .243 WIN, 6MM REM, .257 ROBERTS, .260 REM, 7MM-08 REM, and .308 WIN. There was no boxing around the caliber designations. Some of these magazines, destined for the European market, were made in a two shot version.

The eighth magazine is the current production version and represented the first major change in the design of the magazine since its introduction in 1952. The magazine was redesigned and had a separate matte finished black plastic floor plate. It had the same calibers, now stamped, on its side as the previous version. These magazines fit all autoloader and pump action models chambered for the listed cartridges.

Eighth version magazine with multiple calibers stamped on the side. This current production magazine, introduced in late 2000, had a separate matte finished black plastic floor plate. *Photo: Author. Author's Collection.*

Chapter 14 — **Patents**

The patent markings found on the barrels of the Models 740, 742 and 760 represent the changing views of the Remington Patent Attorney and the Sales Department. The Patent Attorney preferred listing all patents on the barrel to guard against infringement, while the Sales Department's concern was about a cluttered appearance.

The original Model 740 and 760 patent markings were:
PATENT NO. 2,473,373 OTHERS PENDING

This was changed on Sept. 21, 1956 to:
PATENT No. 2,685,754 OTHERS PENDING

The OTHERS PENDING was later changed to AND OTHERS.

The switch in patent

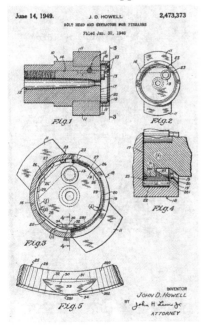

Patent 2,473,373, filed January 30, 1946, issued June 14, 1949 to John D. Howell and assigned to Remington. Covers the bolt head and extractor used on Remington's post war high power rifles including the Model 760.

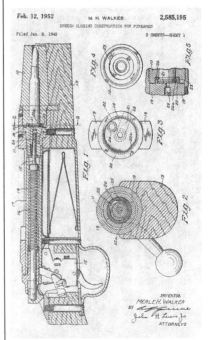

Patent 2,585,195, filed January 8, 1949, issued February 12, 1952 to Merle H. Walker and assigned to Remington. Covers the concept of three rings of steel surrounding the cartridge head.

numbers occurred because No. 2,473,373 did not cover the bolt face and extractor used in the Model 740 rifle.

It appears that the Patent Attorney's views prevailed in 1967 to1968, as the patent markings were expanded to: PATENT NO. 2,685,754 – 2,585,195

2,747,313 – 2,675,638 – 2,473,373 AND OTHERS

The patent markings reverted to a single patent in the mid-1970s.
MADE IN U.S.A. PAT. NUMBER - 2,747,313

The Model 7400 and Model 7600 rifles dropped any reference to the patents.

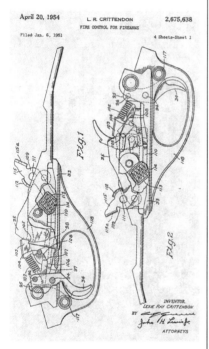

Patent 2,675,638, filed January 6, 1951, issued April 20, 1954 to L. Ray Crittendon and assigned to Remington. Covers the fire control system for the "family of guns".

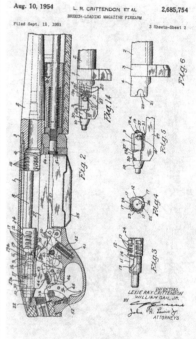

Patent 2,685,754, filed September 12, 1951, issued August 10, 1954 to L. Ray Crittendon and William Gail Jr. and assigned to Remington. Covers the Models 740 and 760.

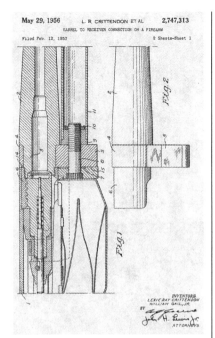

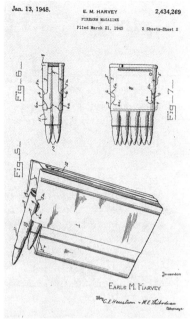

Patent 2.747,313, filed February 12, 1953, issued April 29, 1956 to L. Ray Crittendon and William Gail Jr. and assigned to Remington. Covers the barrel to receiver connection of the Models 740 and 760.

Patent 2,434,269, filed March 21, 1945, issued January 13, 1948 to Earle M. Harvey. Remington, on March 15, 1957, purchased the rights to this patent. Covers the cartridge shoulder ribs and the dual feed lips used on the revised Models 740 and 760 magazines. The patent number is not stamped on the rifle or magazine.

Chapter 15 — **After Market Accessories**

There are a wide variety of aftermarket accessories available for all models of the Remington autoloader and pump action rifles. The two most common aftermarket purchases are scopes, along with the necessary scope mounts and sling swivels. Other popular accessories include receiver aperture sights, safeties, bipod mounts, accuracy blocks and extended capacity magazines.

The following is only a sampling of the various accessories and their manufacturers.

Scope Mounts

There are two different scope mount base screw sizes and tapped hole spacings used in the Remington autoloader and pump action rifles. It is prudent to make sure the right mount base is purchased, as the two bases, especially if out of the original packaging, look similar except for length. The plug screws used to fill the receiver scope mount tapped holes can be used to fill the tapped holes in the barrel when the front and rear sights are removed.

The earlier Models 740, 742 and 760 use a 6-48 screw size and a 3-1/8 inch front to rear tapped hole spacing.

Current production Model 7400 and Model 7600 rifles use a screw size of 8-40 and the distance between the front and rear tapped holes is 3-9/16

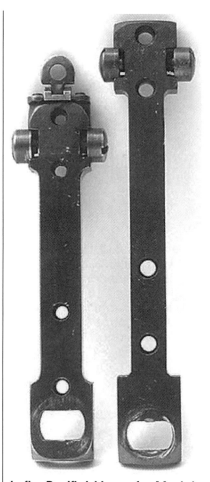

Left – Redfield base for Models 740, 742 and 760. This is an early version with the auxiliary peep sight mounted on the back. Right – Redfield base for Models 7400, 74, Four, 7600, 76 and Six. Note the difference in scope mount base length and screw placement. *Photo: Author. Author's collection.*

inches. The Model Four, Model Six, Sportsman 74 and Sportsman 76 have the same specifications and use the same scope mount base as the Model 7400 and 7600. Remington increased the screw size and lengthened the mount base over the previous models to produce a stronger scope mount base attachment to the rifle.

All current major scope mount manufacturers list a scope mount base for the Model 7400 and Model 7600 rifles. The two major scope mount patterns are

• "Redfield" style with a rotary dovetail front ring and a windage adjustable rear ring

•"Weaver" style cross-slot base.

B-Square, Burris, Leupold, Redfield, Tasco and Weaver are only a few of the current suppliers of one or both of the above patterns for scope rings and bases.

A third scope mounting system allows the use of either iron or scope sights. The Holden Ironsighter and the Williams Sight-Thru are two of the scope mounts that fall into this class.

In addition, there are a number of specialty-mounting systems, such as the Leupold Quick Release Mount System, Tasco World Class Stud System, Weaver pivot rings and the Williams Streamline Top Mount.

Period Scope Mounts

Some Model 740s, 742s and 760s may have unusual scope mounts. All major and many minor scope and scope mount manufacturers from the 1950s to the 1970s made bases for the Remington autoloader and pump action rifles. Some of these manufacturers are still in production. One note of caution though: the earliest Model 760s were not drilled and tapped for scope mounts at the factory, and some were gunsmith drilled and tapped for nonstandard mounts.

A few of the early standard

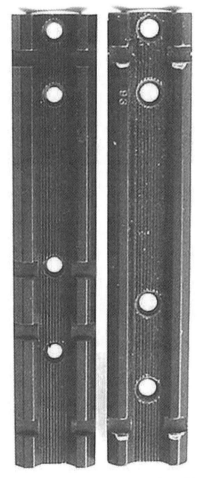

Left – Weaver base number 62 for Models 740, 742 and 760. Right – Weaver base number 93 for Models 7400, 74, Four, 7600, 76 and Six. Note the difference in screw placement. *Photo: Author. Author's collection.*

scope mount bases include:
- Bausch & Lomb
  - Custom two piece base #61-42-59 (not for BDL)
  - Custom one piece base #61-77-10
  - Trophy base #61-76-10
- Buehler
  - Top mount base # 60
- Leupold
  - Adjusto-mount base # 9
  - Model 3 – two piece base # R740

Current "Uncle Mike's" 760 BB quick detachable sling swivel set and full barrel band base fitting Remington Models 740, 742. 7400, 74, Four, 760, 7600, 76 and Six rifles. *Courtesy of Michaels of Oregon.*

Current "Uncle Mike's" 760 ES quick detachable sling swivel set with adapter bolt replacing factory fore-end bolt. Fits all 1968 and later Remington Model 760 pump action rifles. *Courtesy of Michaels of Oregon.*

- Detacho-Mount two piece base # 210 & # 317
- Lyman
  - True-Lock base # B-2
- Marble Arms Corp.
  - See Through base # RE-11-0

Current "Uncle Mike's" 742 ES quick detachable sling swivel set with adapter bolt replacing factory fore-end bolt. Fits all Remington Model 742 ADL or Standard grade autoloader rifles, will not fit Model 742 BDL grade autoloader rifles. *Courtesy of Michaels of Oregon.*

Current "Uncle Mike's" 742 BDL quick detachable sling swivel set with adapter bolt replacing factory fore-end bolt. Fits all Remington Model 742 BDL grade autoloader rifles, will not fit Model 742 ADL or Standard grade autoloader rifles. *Courtesy of Michaels of Oregeon.*

— See Through base # RE-11b-0 (carbine)
• Pachmayr
  – Lo-Swing top mount base # M-740T
  - Lo-Swing side mount base # R740
• United Binocular
  – Two piece base # 160
• Williams
  – Ace in the hole peep top mount base # TM-760
  - Micrometer scope mount base # MIKE -760
  - Offset mount base # 760
  - Quick Convertible side mount base # SM-760

Sling Swivels

Remington installed sling swivels on 1950s and early 1960s ADL and BDL grade rifles and CDL grade carbines. The 1993-1995 Special Purpose Model 7400 and 7600 rifles also were offered with a camouflage Codura sling and sling swivels.

Currently, Michaels of Oregon, under its "Uncle Mikes" brand, offers these sets of quick-detachable sling swivels for the

Current "Uncle Mike's" 7400 quick detachable sling swivel set with adapter bolt replacing factory fore-end bolt. Fits all Remington Model 7400, 74 and Four autoloader rifles. *Courtesy of Michaels of Oregon*

Current Williams Gun Sight Company catalog illustration of their QD-SA-TBM quick detachable sling swivel base barrel adapter on a Remington Autoloader rifle. *Courtesy of Williams Gun Sight Company.*

Remington autoloader and pump action rifles.

760 BB – Models 740, 742, 7400, 74 and Four autoloader rifles, 760 7600, 76 and Six pump action rifles, utilizes a barrel band mount for the front swivel base.

760 ES – 1968 and later Model 760 pump action rifle, adapter bolt with swivel base replaces factory fore-end bolt.

742 ES – Model 742 ADL & Standard grade autoloader rifle, adapter bolt with swivel base, replaces factory fore-end bolt. Does not fit Model 742 BDL grade.

742 BDL – Model 742 BDL grade autoloader rifle only, adapter bolt with swivel base replaces factory fore-end bolt.

1969 Williams Gun Sight Company catalog illustration of their Giant Head Safety. Replaces the standard Remington safety. *Courtesy of Williams Gun Sight Company.*

Does not fit Model 742 ADL or Standard grade.

7400 Four – Model 7400, 74 and Four autoloader rifle, adapter bolt with swivel base replaces factory fore-end bolt.

Williams Gun Sight Company, Inc. offers a universal quick detachable sling swivel set, QD-SA-TBM, which includes a front swivel base that clamps on the barrel.

Rear Sight Base

Model 742 and late Model 760 rear sights have high profiles that can interfere with the front bell of a scope. These screw-attached rear sights are easily removed; however, some hunters prefer to fit an auxiliary rear sight with a fold down leaf as insurance against scope failure. The 6-48 rear sight screw holes are .5625 inches apart. Two suppliers are:

• Lyman Products Company offers the #25B Special Rear Sight Base which allows the use of their #16 Folding Leaf Sight

or any other dovetail folding rear sight.

• Williams Gun Sight Company catalogs the 760 Open Sight Base for any dovetail folding rear sight.

Receiver Sights

Remington drilled and tapped the left receiver panel of early ADL and BDL grade Model 740 and 760 rifles for side-mounted receiver sights. Later rifles dropped the side mounted receiver sight tapped holes.

Lyman Products Corporation advertises the

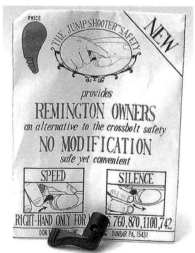

"Jump Shooter Safety" by Don Morrison Inc. A lever action safety replacing the standard Remington safety. It claimed to be faster and quieter than the factory item. *Photo: Author. Author's collection.*

Current Williams Gun Sight Company's top mounted "Guide" receiver sight. *Courtesy of Williams Gun Sight Company.*

following side mounted receiver aperture sights:

- # 66 R - Old Model 740 and 760 rifles.
- #66RH – New Model 740, 742 and 760 rifles

Williams Gun Sight Company Inc. has offered the following side mounted receiver aperture sights, some are still in production:

"Fool Proof" Model

FP-742E - Model 742 below serial number 64,046, Model 760 below serial number 154,965.

FP-760N/740 - Model 740 above serial number 64,046, Model 760 serial number above 154,965.

FP- 740AP - Model 740 above serial number 207,200 in caliber .280 Rem & .30-06, above 200,000 in caliber .308 Win. All Models 742 and 760 with high comb stocks.

#FP-7400 - Models 7400, 74, Four, 7600, 76 and Six, installs without drilling or tapping.

"5-Dollar" Model

5D-760N-740-742 – Model

1969 Williams Gun Sight Company catalog illustration of their Model 740 Accuracy Block. Courtesy of Williams Gun Sight Company.

760s above serial number 154,965, Model 740s above 64,046 and Models 742, 7400, 7600, Four and Six.

Williams currently offers a top mounted receiver sight utilizing existing rear scope mount base screw holes in the top of the receiver.

"Guide" Receiver Sight

WGRS - 7400 Models

Left: Colyer Clip eight cartridge extended capacity steel magazine first offered in the 1960's. Current production Triple K magazines are similar but hold ten cartridges. Right: Colyer Clip box. Photo: Author. Author's collection.

7400, 7600, Four, Six, Sportsman 74 and 76.

WGRS – 742 Models 740, 742 and 760, early models need a higher front sight.

Safeties

Remington has offered a left-handed version of the Model 742 BDL and Model 760 BDL. The safety and Monte Carlo cheek piece stock were left-handed, while cartridge ejection was right-handed. The trigger plate was changed to accommodate a left-handed safety.

Williams Gun Sight Company has advertised Giant Head safeties, both right and left handed, that fit most post war Remington autoloader and pump action rifles and shotguns.

United Sports Manufacturing and Developing Co. of Kensington, Pennsylvania offered the "Magnum Head Safety" for the Model 740 and Model 760 as well as a number of other Remington firearms.

One of the most unusual aftermarket safeties was the "Jump Shooter Safety" by Don Morrison Inc. It was a lever action safety, which replaced the standard Remington cross bolt safety. It was claimed to be faster and, because of the plastic construction, quieter in operation.

Accuracy Block

Williams Gun Company advertised an aluminum block, to be installed in the fore-end of Model 740s, which was claimed to improve accuracy. The Model "S" was for Model 740s under serial number 159,058 and the Model "L" for those above.

Bipod Mount

Harris Engineering has an

U.S.A. Magazine Company current production ten cartridge extended capacity steel magazine. *Photo: Author. Author's collection.*

adapter replacing the fore-end bolt and cap on Model 7400 and Sportsman 74 rifles to allow the use of the Harris Bipod.

Synthetic Stocks

Remington has offered a synthetic stocked version of the Model 7400 and Model 7600 since 1998. The following after market suppliers have advertised a variety of synthetic stocks over the years:

• Ram-Line offers stocks for the Model 7400 and Model 7600.

• Bell and Carlson advertises stocks for the Model 7400, Model 742, Model 760 and Model 7600.

• Outers offers Cadet Gun Stocks designed for the smaller frame shooter. Their stocks fit the Models 740, 742, 7400 and Models 760 and 7600.

Magazines

Remington does not manufacture or sell magazines with a capacity of more than four rounds, so the urge for firepower has been met by at least six

114

different manufacturers. All of the following will function in either the autoloader or the pump action rifle, but will not hold the bolt open after the last shot in the autoloader. One note of caution: game laws in many states limit the number of cartridges to five in the magazine when hunting.

Colyer Clip

The first of the larger capacity magazines, the Colyer Clip appeared in the 1960's. This steel magazine, with a plastic follower, held eight rounds in two columns. The plastic follower sometimes warps from the pressure of the magazine spring. The magazine leaf spring arrangement allows it to be the shortest of the larger capacity magazines.

It was available in two sizes, long for cartridges in the .30-06 class, and short with a spacer for .308 Winchester length cartridges.

Triple K Magazines

This is a current production steel magazine that holds ten rounds in two columns. It is available in two sizes, long for cartridges in the .30-06 class, and short with a spacer for .308 Winchester length cartridges. The bottom of the magazine is the same shape as the Colyer Clip.

USA Magazines

The current production steel magazine from USA Magazines holds ten rounds in two columns and is available in two sizes — long for .30-06 length cartridges, and short with a spacer for .308 Winchester length cartridges. The follower is made of a black polymer and a heavy coiled magazine spring is used.

Eagle

First offered in the mid-1960s, this ten shot single column magazine was made of a translucent high impact polymer. The magazine was claimed to have a constant lifting force for all cartridges due to a pair of volute springs acting on the follower. It was available in three sizes - long for .30-06 class cartridges, 6mm Remington, and short for .308 Winchester class cartridges.

Millett

This current production magazine holds ten rounds in a single column. It is made of translucent Lexan and is available in two sizes — long for

Eagle extended capacity translucent polymer magazine first offered in the 1960's. It holds ten cartridges in a single column. Current production Millett magazines are similar. *Photo: Author. Author's collection.*

.30-06 length cartridges and short for .308 Winchester length cartridges. It is similar in design to the Eagle magazine.

Ramline

This ten shot single column magazine was introduced in the early 1960s. The opaque, black polymer material was advertised not to rust or crack. It was available in two sizes — long for cartridges of .30-06 length and short for .308 Winchester class cartridges. An extended magazine release was included because the Remington factory magazine release was difficult to operate with the magazine in place. A coiled magazine spring was used.

Ramline extended capacity black polymer magazine first offered in the 1960's. It holds ten cartridges in a single column. *Photo: Author. Author's collection.*

Chapter 16 —
Collecting Autoloader and Pump Action Rifles

Remington postwar autoloader and pump action centerfire rifles are appropriate as single pieces in a much larger Remington theme collection. They are also interesting to display as a group by themselves. Approximately four million autoloader and pump action rifles have been sold since 1952, so most of the various models are available today at reasonable prices.

Factory engraved rifles, especially the "F" Premier grade with gold inlays, are at the top of most collectors' want lists. Only 192 Model 740s, 742s and 760s were factory engraved from 1952 to 1980. A few, equally as spectacular, were done outside the factory. These do not command the same premium as factory engraved rifles and the prospective buyer should review the High Grade Rifles chapter for characteristics of Remington factory engraved rifles before the purchase of any engraved rifle.

Commemorative rifles are another area of collector interest. Prices for most commemorative rifles command a premium only if mint in the box. Many of these rifles have been used, and prices should be approximately the same as any used autoloader or pump action rifle.

The etched and gold-plated commemoratives are attractive pieces for a collection and an affordable substitute for a factory engraved "F" grade with gold inlays. The Autoloader 75th Anniversary model is one of the most elaborate commemoratives made to date with its etched and gold plated scenes. It is also the only commemorative Model Four. The 100 Autoloader 75th Anniversary Commemoratives, built on the Model 7400 autoloader rifle, are a distinct variation.

The privately commissioned, non-cataloged Larry E. Benoit Model 7600 Carbine, with his etched and gold-plated signature and deer mount, was the first carbine based commemorative. Only 170 have been made.

The most difficult commemorative to obtain is the engraved and gold embellished 180th Anniversary five gun set, including a Model 7400 and a Model 7600. Fewer than 30 sets were made.

The 150th Anniversary and the US Bicentennial commemorative models were produced in substantial numbers and are available. The Canadian Centennial Model 742 is seldom seen in the US as most of them were sold in Canada in 1967. A number were sold as part of a cased set with a similarly embellished Ruger 10/22 rifle in .22 long rifle caliber.

The 175th Anniversary models offer three variations — the cataloged Model 7400 in caliber .30-06, and the non-cataloged Model 7400 in caliber .270 Winchester and Model 7600 in 7mm-08. Both non-cataloged variations are low production items.

The Model 7600 Grice Deer Hunter in all three calibers and the Buckmaster ADF Model 7400 and Model 7600 represent the first commemoratives commissioned by an outside party and all were limited production items.

A first year production Model 760, especially in caliber .30-06, is becoming harder to find in pristine condition. Many have been drilled and tapped for scope mounts and/or have had sling swivels added. The .300 Savage and .35 Remington chamberings seem to have fewer alterations. These first-year rifles have a YY barrel year code, bright steel ejection port covers, coin slots in the action tube cap, and are not drilled and tapped for scope mounts or receiver sights.

The 1953-1963 ADL and BDL grade Model 740s, 742s, and 760s have been gaining in collector interest. They are a distinct contrast to the rather plain A grade. As previously noted, the only difference between the ADL and BDL grades was the quality of the wood. Only a few early BDL grade rifles had the grade engraved after the model number.

Any premium asked for a BDL grade rifle should be accompanied by documentation as many observed ADL grade rifles have nicely figured wood. As previously noted in the Model 740 and Model 760 chapters, flawed BDL grade figured wood was often downgraded to ADL grade.

The later 1968–1980 BDL Deluxe/Custom Deluxe grades with stepped receivers and pressed basket weave checkering present a different appearance. A left handed option was offered, however, ejection was still right handed.

Model 760s chambered for calibers such as the .222 Remington, .223 Remington, .244 Remington and .257 Remington can command substantial premiums, which vary by region in the US. The 1953-1963 ADL/BDL grade versions in these calibers, as well as in the .280 Remington and .35 Remington, are low production items and are even more desirable.

The Model 740 ADL/BDL grade in .244 Remington and the Model 742 ADL/BDL grade in .280 Remington and 6 mm Remington are also low production items.

The carbines are a group that collectors are just beginning recognize. Any 1960-1963 CDL grade Model 760 is scarce. In fact, any Model 760 carbines in caliber .270 Winchester or .280 Remington are hard to find. The Model 742 carbine in .280 Remington is scarce in either C or CDL grade. The Model 742 CDL grade carbines have game scenes roll marked on both side panels.

The later 1964-1967 and 1968-1980 Standard grade carbines were produced in substantial numbers. The sleepers in the group are the 456 caliber .35 Remington Model 760 carbines produced in the 1960's.

The Model Four and Model Six, intended as replacements for the BDL Custom Deluxe Model 742 and Model 760, were in production from 1981 to 1987. The 6mm Remington was the low production caliber for both models with about 1,500 Model Fours and less than 1,000 Model Sixes made. A Model Four marked .280 Remington instead of 7mm Express Remington has not yet turned up.

The Model 7400 and Model 7600 are still in the line and very little production information is available. Currently, collectors and shooters seem to prize the out of production caliber .35 Whelen in both models.

A wide variety of variations in both the Model 7400 and Model 7600 are not cataloged, but instead, are special ordered by distributors. Grice Wholesale has offered non-cataloged Model 7600 from the 1990 caliber 7mm-08 rifle to the 2001 limited production caliber .35 Remington synthetic carbine and rifle. This "niche" marketing has resulted in a multitude of low production variations about which little is published.

In addition to the Grice special orders the author has noted a number of other non-cataloged rifles and carbines offered for sale on the internet and in trade papers such as the Gun List and Shotgun News.

Very few European market specification Model 7400 and 7600 rifles are available in the United States. These have 2 shot magazines and are in calibers .243, .270, 7mm Exp Rem, .30-06 and .35 Whelen. They have been made with wood or synthetic stocks and with and without Monte Carlo style buttstocks.

The author has a fondness for early production rifles and unusual variations. In the latter category are Model 760s with aluminum receivers, of which at least six were made and two are known to exist. Factory cutaways also are of interest. Eight Model 742s, along with others, were made as demonstrators for regional offices in 1968. Two factory model stamping errors, Model 742s stamped Model 760 *Gamemaster*, have observed.

One factory assembly error, a Model 742 receiver assembled with Model 760 parts, has been reported. One note of caution any rifle claimed to be an assembly error must be accompanied by the original box and intact end label. The label must be stamped Model 760 with the same serial number as on the rifle. A factory caliber marking error, 6 MM REM MAG with the MAG exed out, has been observed on two Model 742s. It is possible that a dual caliber stamping, 7 MM EXP REM / .280 REM may have been stamped on Model 7400s and Model 7600s but so far the author is not aware of any rifles so marked.

The author purchased a non-cataloged enhanced receiver Model 7600 in .25-06 Remington with the intention of keeping it as a mint in box collector's item. Good friend Ray Shields kept urging that it should be shot to see if it was as accurate as his Remington pumps. He even offered use of his loading dies and a selection of bullets. A 4x – 16x variable scope intended for a varmint rifle was installed and sighted in at the local range. It quickly became apparent when we moved back to the 100 yard range that the rifle was capable of good groups as the first three shots covered 1-1/2 inches. The second and third set of three shots went in a 1-1/4 inch group.

The next session with carefully loaded ammunition and proper benchrest technique regularly produced three shot groups of one inch or less. The best group, with 100 grain Nosler Ballistic Tip bullets, was just over three-quarters of an inch.

The author does not remember the name of the outdoors magazine or the writer who long ago suggested that the best measure of a hunting rifle's accuracy are the first three shots out of a cold barrel. The first three shots grouping in an inch at 100 yards has been this author's standard for many years. This Model 7600 is a keeper as it consistently places the first three shots in an inch or less, right on target.

Remington's autoloader, often called the Maine woods rifle, and pump action rifles, long a favorite in Pennsylvania, are known as hunter's guns. They are carried in the woods under all sorts of conditions and their reliability is legendary. Many early models are still in use today.

One final story: the author was informed about a caliber .280 Remington Model 760 CDL grade carbine — a rare variation. The owner, about the same age as the author, was contacted, information gathered and in the course of conversation was asked about selling it. He said he had absolutely no interest in selling his favorite carbine. This was his first new rifle, and since 1962 had been carried on many successful hunting trips. While the gentleman now has a number of newer hunting rifles, this carbine is an old friend that had never let him down. This feeling of loyalty has turned up in a number of other interviews.

Chapter 17 — **Serial Numbers and Barrel Codes**

Each modern Remington firearm carries production information on its frame and barrel. Deciphering that information will indicate when the firearm was made and whether or not it was returned to the factory for repairs.

Serial Number

The serial number is stamped on the left receiver panel, under the *Remington* logo. Later production rifles dropped the line under *Remington*. The name *Gamemaster* or *Woodmaster* was roll stamped above the model designation between the fire control pins. The names were dropped when the Model 7400 and Model 7600 were introduced.

The production Model 740, 742 and 760 each started at serial number 1001. Lower serial numbered guns were experimental or preproduction rifles. The Model 740 production ended in 1960 at serial number 253821. The Model 742 and 760 serial number sequence changed on November 26, 1968 as a result of the 1968 Gun Control Act, which required that no two guns by the same manufacturer have the same serial number. The initial serial number sequence of the Model 742 ended at 396562, while the initial serial number sequence of the Model 760 ended at 549773.

Both models began sharing a new block of serial numbers after November 26, 1968. This block began at A6900000 and ended at A7499000. The second block of shared serial numbers began January 3, 1978 at B6900000 and ended at B7458723. The Model Four and Model Six serial numbers started at A4000000 in late 1980. The Model 7400 and Model 7600

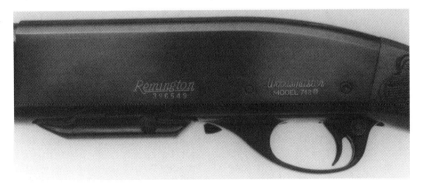

Left receiver panel with *Remington* logo, serial number, model name and number. Later production rifles drop the line under *Remington*. Photo: Author, Author's Collection.

initial serial numbers started at 8000000 in late 1980, began a second block starting at A8000000 in 1989, and a third block starting at B8000000 in 1991.

A duplicate serial number will have a Z stamped after the serial number. This allowed Remington to comply with the federal law without scrapping a usable receiver.

High grade rifles produced by the Remington Custom Shop were not roll marked with the *Remington* logo, model designation or serial number. The bottom of the barrel extension was hand engraved with *Remington,* while either the bottom or the left panel of the receiver was hand engraved with the model number, grade and serial number.

Barrel Code Stamping

Barrel code stampings provided Remington with a record of production for service and repair. The right side of the barrel had the following marks from left to right: REP in an oval (Remington English Proof), letter (heat treat), test character (function) and target character (accuracy). The left side of the barrel had the magnaflux stamp about six inches from the receiver near the rear sight.

Additional stamps appear about 1-1/2 inches from the receiver, from left to right — final inspector's character, date code (month, year letters), and the assembler's letter or number code. If the rifle had been returned to the factory for service it will have a date code (month, year letters), 3 (customer repair) and the customer repair

inspector character stamped ahead of the original final inspector character.

Each Remington inspector had his own unique character stamp. These can take the shape of an arrow, cross, heart, star,

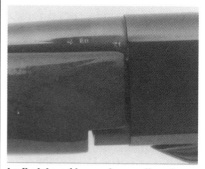

Left side of barrel, reading from left to right – Final inspector's character, date code (month, year letters), assembler's number code. The magnaflux stamp, a triangle with letter enclosed, not shown, is just after the barrel legend. This rifle has the date code EO indicating assembly October 1977. The fore-end is moved slightly forward. *Photo: Author, Author's Collection.*

Right side of barrel, reading from left to right: REP (Remington English Proof), letter (heat treat), test character (function), target character (accuracy). Fore-end is moved slightly forward. *Photo: Author, Author's collection.*

trefoil or other shapes. High grade rifles may not be stamped or may be stamped only on the bottom of the barrel covered by the fore-end.

From first production of these rifles, Remington considered the fire control system so crucial that assemblers stamped their letter or number code on the right side of the trigger plate.

The barrel date code provides the month and year that the rifle was assembled and sent to the warehouse. The first letter was the month according to the following code:

B – January
L – February
A – March
C – April
K – May
P – June
O – July
W- August
D – September
E – October
R – November
X – December

The year code was the second and, in the case of the early 1950s, third letter. The letters were individually stamped in the early years, resulting in an uneven spacing. A jig was used later to provide even spacing, but also resulted in lightly struck date codes.

Year Codes
1950 WW
1951 XX
1952 YY
1953 ZZ
1954 A
1955 B
1956 C
1957 D
1958 E

1959 F
1960 G
1961 H
1962 J
1963 K
1964 L
1965 M
1966 N
1967 P
1968 R
1969 S
1970 T
1971 U
1972 W
1973 X
1974 Y
1975 Z
1976 I
1977 O
1978 Q
1979 V
1980 A
1981 B
1982 C
1983 D
1984 E
1985 F
1986 G
1987 H
1988 I
1989 J
1990 K
1991 L
1992 M
1993 N
1994 O
1995 P
1996 Q
1997 R
1998 S
1999 T
2000 U

The date code has to be examined in conjunction with the production years of the model. A Model 760 with a XA code could be either an early production December 1954 or a late production December 1980

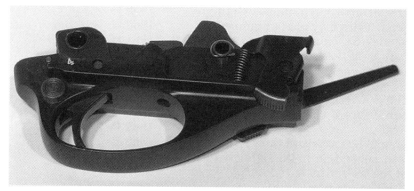

Assembler's code (4) stamped on the right side of a Model 760 pump action rifle trigger plate. *Photo: Author, Author's collection.*

rifle. The stock, sights and other characteristics would have to be examined to determine which year it was made.

Other year date code overlaps are easier to resolve by using the model designation. A Model 740 (produced from 1954 to 1959) can have a year code from A to F. A Model 742 (produced in 1980 or 1981) would be year coded A or B and a Model 7400 (produced from 1982 to 1985) would be year coded C to F.

On August 9, 1999 Remington stopped stamping the month and year date code on the barrel. but continued to stamp it on the shipping box end label. The end label is of significance as it provides a description of the rifle, serial number, order number and packer code. The four digit order number is preceded by a 2 (a computer classification number) on later shipping box labels. The packer code is found usually on the lower right hand corner and consists of the packer's numerical code and the letter date codes.

In the 1990s the packer's code, a numerical and/or letter combination, was moved from the lower right hand corner and is now stamped in smaller type to the left of the order number. The date code is still stamped in the lower right hand corner and now includes the actual day of assembly.

One word of caution regarding date codes: these represent the assembly date of the rifle and not the shipping date from the warehouse. Slow-selling calibers may languish in the Remington warehouse for several years after being dropped from production. The 1956 Remington catalog noted that the Model 760s in caliber .257 Roberts were "subject to stock on hand." The 1958 catalog noted the Model 760 in caliber .300 Savage, as well as the .257 Roberts, also were "subject to stock on hand". The 1960 catalog added Model 760 in .222 Remington and .244 Remington to the list of calibers "subject to stock on hand."

Generally, Remington introduced new models and

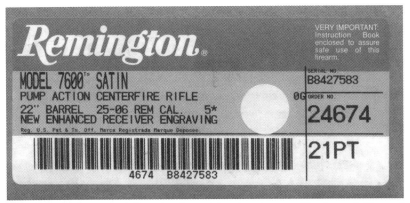

End label of current Remington shipping box with description of rifle, serial number, order number, packer's code and date code. The packer code, a numerical and letter code (0G), is now in smaller letters and has been moved to the left of the order number box. The lower right hand corner has the date code 21PT indicating assembly 21 June, 1999. *Photo: Author, Author's collection.*

variations at their winter sporting press seminar held in December of each year. This was followed by a catalog introduction to the public in January or February. Some calibers, especially in the early years, were introduced in mid-year. These calibers and dates are noted in each model's chapter.

In this book, the author will normally use the catalog or price list date as the date of introduction for changes. It should be noted that Remington used up surplus parts before switching to a new style. This is true of ejection port covers, buttplates, sights, and stocks as well as internal parts.

The catalogs, useful as a reference for certain items, do have a few inconsistencies. Stock dimensions vary from catalog to catalog, especially in the 1960-to-1975 and 1987-to-1989 periods. Measurements of rifles made during these periods do not reflect the changes reported in the catalog.

Chapter 18 —
Acknowledgments

I wish to thank all the collectors and dealers who, over the years, have taken time to fill out surveys and answer questions about their rifles. This volume would be much slimmer without their help.

A special thank you goes to a fine gentleman who took some time with a youthful collector many years ago. Jerry Crozier of Crozier's Gunroom, Homer, NY introduced a college student to the fascinating history behind that old rifle in the rack. I was hooked and began a most enjoyable hobby.

Ray Shields, close friend and neighbor, asked a question about his old Model 760 rifle that started the author on a quest, resulting in this book.

Jack Heath, Remington Arms Company Historian offered many helpful suggestions in reviewing the original draft. He also shared his knowledge, as well as many photographs, catalogs, product announcements and price lists that he had filed away during his long career as Manager, Customer Information with the Remington Arms Company. Jack also serves as Vice President of the Remington Society of America.

Roy Marcot, editor of the Remington Society Journal and member of RSA Research Team, provided some of the photographs and shared his files. He also introduced the author to the RSA Research Team. The information on the early development of the autoloader and pump action rifles discovered in the Remington Arms Company Archives is a direct result of his assistance.

Jim Hennings provided information on prototype firearms in the Remington Research and Development Gun Library and production data on discontinued models.

Dennis Sanita and Fred Supry assisted the Remington Society Research Team and the author in innumerable ways during visits to the Remington Arms Co. factory in Ilion, NY.

Jerry Beigh and Lyle Wheelock graciously shared their office with the Remington Society Research Team and the author during visits to the Remington Arms Co. factory.

Jim Martin provided information on the Model 742X break open project as well as developmental details on the Models 7400 and 7600. He was the Section Manager in charge of both projects until his retirement from Remington Arms Co. in 1983.

Sam Alvis as Manager, Arms Research & Development Division, Remington Arms Company and later Curator of the Remington Arms Company Museum, preserved many records of firearms development during the 1950s and 1960s. His foresight was of inestimable value to present and future researchers.

John Lacy, RSA member and author of *Remington 700 25 Years 1962 – 1987*, provided information on the factory cutaway rifles and the Model 742s marked 6MM XXX.

Robert Creamer, Remington Society of America treasurer and member of the RSA Research Team, provided information on the Model Four Limited Edition rifles including the late production variation.

Jim Eckley, General Manager of Grice Wholesale, provided information on the Grice non-cataloged Model 7600 variations.

Larry "Babe" Del Grego graciously allowed us to review Robert P. Runge's engraver's "pulls" and work order tags in his collection.

Richard Pearson of the "Clip Joint" reviewed the chapter on magazines and provided some of the less common magazines illustrated.

Steve Scribner, Wilderness Trading and Supply Co., gave us the opportunity to photograph the Larry Benoit Commemorative Carbine.

George Coldren provided information on his Model 742 assembled with Model 760 parts.

Wildwood Inc. provided information on its Model 742-stamped Model 760.

Blair Grabski provided information and photographs of the 180th Anniversary Rifles.

Williams Gun Sight Company provided copies of its early catalogs.

Friend, editor and publisher, Alan M. Petrillo, encouraged the author to prepare the manuscript, and then tackled the publishing job with great enthusiasm and skill. He knew in advance the difficulty of his task since he edited and published my previous two books — *The Remington-Lee Rifle* and *The Winchester-Lee Rifle*.

In addition to those mentioned above, the author thanks the following for assistance of various kinds:
Bill Appleton
Tom Cameron
Gordon Fosburg
Roger Leveiller
Art Lewis
Richard Van Duesen Jr.

A special tribute goes to past and present Remington Arms Company designers and craftsman who have developed and produced firearms of distinction. The fruits of their labors have provided much satisfaction to owners and users over the years. It is they who made this book possible in the first place.

127

Chapter 19 —
Bibliography

Articles

Carpenter, Donald F.; "Notebook – Remington Arms Production – World War II," Excerpts published in *U.S. Martial Arms Collector*, Volume 92.

Cotterman, Dan; "Remington's Model 760: This is known as a butterfly gun; with changee-changee barrels for numerous calibers!" *Gun World*, January 1960.

McCormack, Kevin; "When I Paint My Masterpiece. The Life and Times Of Robert P. Runge, Parker Master Engraver," *The Double Gun Journal*, Autumn 2000.

Petrini, Frank B.; "The Models Four and Six – Remington's Refined Centerfires," *Shooting Times*, September 1981.

Rees, Clair; "Slide-Action Deer Guns," *Guns & Ammo*, February 2001.

Sundra, Jon R.; "Remington's Slide Action M760 Rifle," *Shooting Times*, February 1977.

Towsley, Bryce M.; "No Bolt Deer Guns," *American Rifleman*, August 1999.

Books

Benoit, Larry; *The Beginning: Where It All Began*, Benoit Enterprises, 1992.

Fjestad, S. P.; *Blue Book of Gun Values*, Blue Book Publications, various from 1990.

Hatch, Alden; *Remington Arms in American History*, Rhinehart & Co. 1956.

Lacy, John F.; *The Remington 700: A History and Users Manual, 1962 – 1987*. John F. Lacy, 1989.

Marcot, Roy; *Remington "America's Oldest Gunmaker,"* Primedia Special Interest Publications, 1998.

Myszkowski, Eugene J.; *The Remington-Lee Rifle*, Excalibur Publications, 1994.

Stroebel, Nick; *Old Gunsights, a Collector's Guide 1850 - 1965*, Krause Publications, 1998.

Stroebel, Nick; *Old Scopes*, Krause Publications, 2000.

Towsley, Bryce M.; *Big Bucks the Benoit Way: Secrets From America's First Family of White Tail Hunting*, Krause Publications, 1998.

Catalogs

Annual Spring Firearms Auction - The Jack Appel Collection of Remington Firearms. Richard W. Oliver, Auctioneer, Kennebunk, Maine, May 20, 1989.

Public Auction – Antique

Weapons – Collection of John T. Amber. Richard A. Bourne, Auctioneer, Hyannis, Massachusetts, November 11-13, 1986.

The Guns of Remington; A catalog of the exhibition *It Never Failed Me: The Arms & Art of the Remington Arms Company*, Buffalo Bill Historical Center, Cody Wyoming, Biplane Productions, 1997.

B-Square Company; various.

Blount Inc., Weaver/Ramline, various.

Leupold & Stevens, various.

Lyman Products Corporation, various.

Williams Gun Sight Company, various.

Remington Arms Company Announcements, Catalogs and Reports
Remington Arms Company Annual Catalogs 1952 – 2001.

Remington Arms Company Annual Price Lists – various from 1952.

Remington Arms Company Annual News Release Kits - various from 1980.

Remington Arms Company Limited Edition News Release – 1982.

Remington Arms Company Custom Shop Brochures – various from 1994.

Remington Arms Company Product Promotion Kits - various from 1982.

Remington Arms Company Model Announcement Sheets – various from 1952.

Remington Arms Company Instruction Folders – various from 1952.

Remington Arms Company Product Committee Meeting Reports – 1935 to 1945.

Remington Arms Company internal committee reports and memorandums, order books, photographs, shipping records, work books, etc. Remington Archives, Remington Arms Company Inc. Ilion, New York.

Chapter 20 — **Index**

It is recommended that Chapter 17, Serial Numbers and Barrel Codes, be reviewed first to determine the year a rifle was assembled.

Page numbers in **bold** are photographs or illustrations.

Models Four and Six Autoloader and Pump Action Rifles

Model 49 prewar pump action rifle prototype – 4

Model 740 Autoloader Rifle

Model 742 Autoloader Rifle

Model 760 Pump Action Rifle

130

Chapter 21 —

About the Author

The author was raised on the family dairy farm some 15 miles from Ilion, NY. Many years ago, a gun dealer answered a "what is that gun?" question with a quick history lesson that began a lifelong hobby.

Myszkowski is the author of **The Remington-Lee Rifle** and **The Winchester-Lee Rifle**, also published by Excalibur Publications.

He has also published articles in *American Riflemen, Gun Report, Shooting Times*, the Remington Society *Journal* and the Dixie Gun Works catalog.

Myszkowski, a member of the Remington Society of America since 1982, currently serves as one of its directors, and is a member of the Research Team. He has been a member of the National Rifle Association since 1962.

He is retired from New York State government service and lives with his wife, Jan, in Stuyvesant Falls, NY.

The author welcomes correspondence concerning Remington postwar center fire autoloader and pump action rifles. It can be sent to him at P.O. Box 167, Stuyvesant Falls, NY 12174 or at genejan61@berk.com.